AF706982

Societal Responsibility of Artificial Intelligence

Technological Prospects and Social Applications Set

coordinated by
Bruno Salgues

Volume 4

Societal Responsibility of Artificial Intelligence

Towards an Ethical and Eco-responsible AI

Jérôme Béranger

iSTE

WILEY

First published 2021 in Great Britain and the United States by ISTE Ltd and John Wiley & Sons, Inc.

Apart from any fair dealing for the purposes of research or private study, or criticism or review, as permitted under the Copyright, Designs and Patents Act 1988, this publication may only be reproduced, stored or transmitted, in any form or by any means, with the prior permission in writing of the publishers, or in the case of reprographic reproduction in accordance with the terms and licenses issued by the CLA. Enquiries concerning reproduction outside these terms should be sent to the publishers at the undermentioned address:

ISTE Ltd
27-37 St George's Road
London SW19 4EU
UK

www.iste.co.uk

John Wiley & Sons, Inc.
111 River Street
Hoboken, NJ 07030
USA

www.wiley.com

© ISTE Ltd 2021
The rights of Jérôme Béranger to be identified as the author of this work have been asserted by him in accordance with the Copyright, Designs and Patents Act 1988.

Library of Congress Control Number: 2021931051

British Library Cataloguing-in-Publication Data
A CIP record for this book is available from the British Library
ISBN 978-1-78630-694-4

MIX
Paper from responsible sources
FSC® C013604

Contents

Acknowledgments

This book is first and foremost the story of a professional human adventure that has been the fruit of much support and exciting interactions over the last few years.

In particular, I wish to thank the UMR 1295 BIOETHICS team of the Inserm/Université Paul Sabatier de Toulouse mixed research unit, and the UNESCO chair on "Ethics, Science, and Society" of the Université Fédérale de Toulouse for their trust and support in my research work on the ethical approach to AI.

Without all the people I met, this book would never have seen the light of day! I thank them very much.

Introduction

To the great promises that AI brings, it must meet a great human responsibility.

The digital world is characterized by its immediacy, its density of information, its omnipresence, in contrast to the concrete world of things. Now, with the multiplication of means of connection, the decrease in technology costs, the new capacities of data collection and algorithmic processing, we realize that we can communicate elements of our environment that were silent until now. We are witnessing the multifaceted development of new information and communication technologies (NICTs), illustrated by the emergence of technologies associated with Big Data; connected objects; algorithms; nanotechnology, biotechnology, information technology, and cognitive science (NBIC); blockchain; artificial intelligence (AI); virtual and augmented reality; and even quantum computing. AI is developing at an extremely rapid pace. We should expect to see significant changes in our society as AI systems become embedded in many aspects of our lives.

This multifaceted digital phenomenon is bringing different universes together by adding the speed, intelligence, and ubiquity of digital technology to the objects associated with these NICTs. Major developments related to AI in healthcare, autonomous vehicles, cybersecurity, education, home and service robots are improving the quality and comfort of our lives every day. Now, AI is fundamental to address many of the major challenges facing humanity, such as climate change, global health and well-being, natural resource development, and reliable and sustainable legal and democratic systems. This technology is, therefore, changing the way we live, consume, function and work. This is illustrated by a disruption with the past in the relationship and link that each person has with his or her neighbors. From then on, these interactions force the system to rethink each human activity. This is the beginning of a silent but very present revolution that is happening right before our eyes. A new era of change and disruption where survival inevitably requires reactivity, adaptability, creativity and, therefore, innovation.

Consequently, this technoscientific context is conducive to the development of an increasingly important international cultural and intellectual movement, namely transhumanism, whose objective is to improve the physical and mental characteristics of human beings by relying on biotechnologies and other emerging technologies. This current of thought considers that certain states of the human condition such as illness, disability, pain, aging and death are not fatal in themselves and can be corrected or even eliminated.

Thus, technological revolutions have enabled a change of scale in the exploitation of digital data, particularly in the field of genetics. They can be produced in large quantities, in an increasingly precise manner and preserved over an indefinite period of time. It can be observed that advances in computer science have made it possible, through the creation of specific programs, for databases to be interoperable, thus allowing for the fusion of data from various and multiple sources. To this, we can add the development of new ways of accessing data, in particular through the multiplication of data sources of all kinds. Crowdsourcing[1] is becoming one of the new devices allowing easy access, in real time, to digital data in order to develop research (Khare *et al.* 2015).

ALGORITHMIC PROCESSING.–

Algorithmic processing is a finite and unambiguous sequence of operations or instructions to solve a problem or obtain a result. Algorithms are found today in many applications such as computer operation, cryptography, information routing, resource planning, and optimal use of resources, image processing, word processing and so on. An algorithm is a general

1 In France, crowdsourcing is defined according to the *Commission générale de terminologie et de néologie* (2014) as the "mode of completion of a project or a product calling for contributions from a large number of people, generally Internet users". *JORF*, 0179(91), 12995.

method for solving a set of problems. It is said to be correct when, for each instance of the problem, it ends up producing the correct output, i.e. it solves the problem.

BIG DATA.–

Big Data, or megadata, sometimes referred to as massive data, refers to data sets that become so large that they are difficult to make use of with traditional database or information management tools. The term Big Data refers to a new discipline at the crossroads of several sectors such as technology, statistics, databases and business (marketing, finance, health, human resources, etc.). This phenomenon can be defined according to seven characteristics, the 7Vs (volume, variety, velocity, veracity, visualization, variability, value).

BLOCKCHAIN.–

Computer "block chain" is protected against any modification, each of which contains the identifier of its predecessor. The blockchain records a set of data such as a date, a cryptographic signature associated with the sender and a whole set of other specific elements. All these exchanges can be traced, consulted and downloaded free of charge on the Internet, by anyone who wishes to check the validity and non-falsification of the database in real time. The major advantage of this device is the ability to store a proof of information with each transaction in order to be able to prove later and at any moment the existence and content of this original information at a given moment. Its mission is, therefore, to create trust by protocolizing a digital asset or database by making it auditable.

CROWDSOURCING.–

A practice that corresponds to appealing to the general public or consumers to propose and create elements of the marketing policy (brand choice, slogan creation, video creation, product ideation/co-creation, etc.) or even to carry out marketing services. Within the framework of crowdsourcing, professional or amateur service providers can then be rewarded, remunerated or sometimes only valued when their creations are chosen by the advertiser or sometimes simply for their participation effort. Crowdsourcing has especially developed with the Internet, which favors the soliciting consumers or freelancers through specialized platforms.

AI appears as an essential evolution in the processing of digital information. It represents the part of computing dedicated to the automation of intelligent behaviors. This approach is the search for ways to endow computer systems with intellectual capacities comparable to those of human beings. AI must be capable of learning, adapting and changing its behavior.

The idea of elaborating autonomous machines probably dates back to Greek antiquity with the automatons built by Hephaestus, reported notably in the *Iliad* (Marcinkowski and Wilgaux 2004). For Brian Krzanich, President and CEO of Intel (the world's leading microprocessor manufacturer), AI is not only the next tidal wave in computing, but also the next major turning point in the history of humankind. It does not represent a classic computer program: it is more educated than programmed. It is clear that the AI lawsuit has mixed fantasy, science fiction and long-term futurology, forgetting even the basic definitions of the latter.

Thus, the concept of AI[2] is to develop computer programs capable of performing tasks performed by humans requiring learning, memory organization, and reasoning. The objective is to give notions of rationality, reasoning and perception (e.g. visual) functions to control a robot in an unknown environment. Its popularity is associated with new techniques, such as deep learning, which gives a program the possibility to learn how to represent the world because of a network of virtual neurons that perform each of the elementary calculations, in a similar way to our brain.

> DEEP LEARNING.–
>
> This algorithmic system has been used for more than 20 years for different actions in the form of neural networks, in particular to "learn". A neuron represents a simple function that takes different inputs and calculates its result, which it sends to different outputs. These neurons are mainly structured and organized in layers. The first layer uses almost raw data and the last layer will generate a result. The more layers there are, the greater the learning and performance capacity will be. One can take the example of character recognition from handwriting. The first layer will take into account all the pixels that make up a written character – for example, a

2 ISO 2382-28:1995 defines artificial intelligence as "the capability of a functional unit to perform functions that are generally associated with human intelligence, such as reasoning and learning".

letter or a number – and each neuron will have a few pixels to analyze. The last layer will indicate "it's a T with a probability of 0.8" or "it's an I with a probability of 0.3". A backpropagation operation is performed from the final result to remodify the parameters of each neuron.

The machine is programmed to "learn to learn". AI does not exist to replace people, but to complement, assist, optimize and extend human capabilities. There are two types of AI:

– weak AI: its objective is to rid people of tedious tasks, using a computer program reproducing a specific behavior. This AI is fast to program, very powerful, but without any possibility of evolution. It is the current AI;

– strong AI: its objective is to build increasingly autonomous systems, or algorithms capable of solving problems. It is the most similar approach to human behavior. This AI learns or adapts very easily. Thanks to algorithmic feedback loops, the machine can modify its internal parameters used to manage the representation of each stratum from the representation of the previous stratum. These strata of functionalities are learned by the machine itself and not by humans. From this postulate, we can say that the machine becomes autonomous and intelligent, by constructing its own "computerization" structures and relying on axiomatic decisions. It is the future AI that should be developed in about 10 years.

WEAK AI.–

Weak AI or narrow AI simulates specific cognitive abilities such as natural language comprehension, speech recognition or driving. It only performs tasks for which it is programmed. It is therefore highly specialized. It is a machine for which the physical world is somewhat enigmatic, even ghostly, if it perceives it at all. It does not even have any awareness of time. This AI is unintelligent and works only on the basis of scenarios pre-established by designers and developers.

STRONG AI.–

Artificial general intelligence (AGI) or strong AI has similar – and even superior – reasoning abilities to those of human beings. It is endowed with capabilities not limited to certain areas or tasks. It reproduces or aims to reproduce a mind, or even a consciousness, on a machine. That is to say, an evolutionary machine with its own reasoning and consciousness, capable in particular of independently elaborating strategies and/or decisions that go beyond human beings in order to understand them so as to help them (in the best of cases) or to deceive or even destroy them (in the worst of cases).

From a general point of view, AI can be illustrated as an algorithmic matrix that aims to "justly or coldly" optimize decisions. Naturally, the morality or fairness of this judgment is not predefined, but depends, on the one hand, on the way in which the rules are learned (the objective criterion that has been chosen), and, on the other hand, on the way in which the learning sample has been constructed. The choice of the mathematical rules used to create the model is crucial. Just like the human functioning that analyzes a situation before changing one's behavior, AI allows the machine to learn from its own results to modify its programming. This technology already exists in many applications like on our smartphones, and should soon be extended to all areas of daily life: from medicine to the autonomous car, through artistic creation, mass distribution, or the fight against crime and terrorism. Machine learning not only offers the opportunity to automatically make use of large amounts of data and identify habits in consumer behavior. Now, we can also actuate these data.

MACHINE LEARNING.–

Machine learning concerns the design, analysis, development and implementation of methods that allow a machine (in the broadest sense) to evolve through a systematic process, and, thus, perform tasks that are difficult or impossible to perform by more traditional algorithmic means. The algorithms used allow, to a certain extent, a computer-controlled (possibly a robot) or computer-assisted system to adapt its analyses and response behaviors based on the analysis of empirical data from a database or sensors.

In our view, adopting the machine learning method is no longer just a utility, but rather a necessity. Thus, in light of the digital transition and this "war of intelligences" (Alexandre 2017), companies will be the target of a major transformation and will invest in AI applications in order to:

– increase human expertise via virtual assistance programs;

– optimize certain products and services;

– bring new perspectives in R&D through the evolution of self-learning systems.

Therefore, AI holds great promise, but also strong fears, hazards and dangers that must be corrected or even removed, to ensure an implementation that is in accordance with the legal framework, moral values and ethical principles, and the common good. The conflicts in question can be very varied. Indeed, machines like robotic assistants ultimately ignore the concepts of good and evil. They need to be taught everything. Autonomous cars are likely to involve us in accidents or dangerous situations. Some conversational agents may insult or give bad advice to individuals and not be kind to them.

Thus, even if today, ethical recommendations have little impact on the functional scope of AI and introduce an additional level of complexity in the design of self-learning systems, it becomes essential, in the future, to design and integrate ethical criteria around digital projects related to AI.

Several standards dealing with algorithmic systems, transparency, privacy, confidentiality, impartiality and more generally with the development of ethical systems have been developed by professional associations such as the IEEE (Institute of Electrical and Electronics Engineers) and the IETF (Internet Engineering Task Force)[3].

To this can be added documents focusing on ethical principles related to AI, such as:

– the Asilomar AI Principles, developed at the Future of Life Institute, in collaboration with attendees of the high-level Asilomar conference of January 2017 (hereafter "Asilomar" refers to Asilomar AI Principles, 2017);

– the ethical principles proposed in the Declaration on Artificial Intelligence, Robotics and Autonomous Systems, published by the European Group on Ethics in Science and New Technologies of the European Commission, in March 2018;

3 IEEE P7000: *Model Process for Addressing Ethical Concerns During System Design*; IEEE P7001: *Transparency of Autonomous Systems*; IEEE P7002: *Data Privacy Process*; IEEE P7003: *Algorithmic Bias Considerations*; IETF *Research into Human Rights Protocol Considerations* draft.

– the principles set out by the High-Level Expert Group on AI, via a report entitled "Ethics Guidelines for Trustworthy AI", for the European Commission, December 18, 2018;

– the Montreal Declaration for AI, developed at the University of Montreal, following the Forum on the Socially Responsible Development of AI of November 2017 (hereafter "Montreal" refers to Montreal Declaration, 2017);

– best practices in AI of the *Partnership on AI,* the multi-stakeholder organization – composed of academics, researchers, civil society organizations, companies building and utilizing AI academics, researchers, civil society organizations and companies building and utilizing AI – that, in 2018, studied and formulated best practices in AI technologies. The objective was to improve public understanding of AI and to serve as an open platform for discussion and engagement on AI and its influences on individuals and society;

– the "five fundamental principles for an AI code", proposed in paragraph 417 of the UK House of Lords Artificial Intelligence Committee's report, "AI in the UK: Ready, Willing and Able", published in April 2018 (hereafter "AIUK" refers to House of Lords, 2018);

– the ethical charter drawn up by the European Commission for the Efficiency of Justice (CEPEJ) on the use of AI in judicial systems and their environment. It is the first European text setting out ethical principles relating to the use of AI in judicial systems (see Appendix 1);

– the ethical principles of Luciano Floridi *et al.* in their article entitled "AI4People – An Ethical Framework for a Good AI Society: Opportunities, Risks, Principles, and Recommendations", *Minds and Machines*, December 2018;

– the OPECST (*Office parlementaire d'évaluation des choix scientifiques et technologiques*) report (De Ganay and Gillot 2017);

– the six practical recommendations of the report of the CNIL *(Commission nationale de l'information et des libertés)*[4] on the ethical issues of algorithms and AI, drafted in 2017 (see Appendix 2);

– the report published by the French member of parliament Cédric Villani (2018) on AI;

– the Declaration on Ethics and Data Protection in the Artificial Intelligence Sector, at the 40th International Conference of Data Protection and Privacy Commissioners (ICDPPC), Tuesday, October 23, 2018, in Brussels;

– the seven guidelines[5] developed by the European High Level Expert Group on AI, published on April 8, 2019 by the European Commission;

– the five principles set out in the OECD Council Recommendation on the development, implementation and use of AI, adopted on May 22, 2019, by the Council to OECD Ministers[6].

What is the best practice of ethical frameworks, regulations, technical standards, and best practices that are environmentally sustainable and socially acceptable? It is clear that these shared frameworks do not guarantee success. Mistakes and illegal behavior continue to occur. But their availability requires a clear and precise idea of what needs to be done and how to evaluate competing solutions.

This diversity of approaches and initiatives on the subject reflects the major challenge of establishing a common framework for ethical governance of AI. This raises a delicate and decisive question: how should the ethical governance of AI be defined, or by which characteristics? What are the "measurable" values, translating notions of loyalty, responsibility, trust and

4 CNIL (2017). *Comment permettre à l'homme de garder la main ? Les enjeux éthiques des algorithmes et de l'intelligence artificielle.* Summary report of the public debate led by the CNIL in the context of the mission of ethical reflection entrusted by law for a digital Republic.

5 These seven essential requirements include human factor and human control, technical robustness and security, privacy and data governance, transparency, diversity, non-discrimination and equity, societal and environmental well-being, and accountability.

6 On May 22, 2019, through the OECD Council of Ministers, 42 countries (the 36 OECD countries and Argentina, Brazil, Colombia, Costa Rica, Peru, and Romania) adopted the principles set out in the OECD Recommendation on AI, making it the first intergovernmental agreement to stimulate innovation and build confidence in AI by promoting a responsible approach to trusted AI, while ensuring respect for human rights and democratic values.

thus ethics applied to algorithmic decisions when they are the consequence or the result of a prediction?

It is from this vision of universalization that we felt the need to write this book around the framework of AI applicable to all. As a result, we have developed a moral framework to support digital AI projects by observing a number of requirements, recommendations and rules, elaborated, verified and discussed at each stage of design, implementation and use. This allowed us to design ethical criteria, according to our determinants, both essential and universal, based on the principle of Ethics by Design[7] or Human Rights by Design to move toward a totally innovative principle of Ethics by Evolution that we will develop throughout this book. The objective is to achieve AI that is safer, more secure and better adapted to our needs, both ethical and human, over time. This will help optimize our ability to monitor progress against criteria of sustainability and social cohesion. AI is, therefore, not an end in itself, but rather a means to increase individual and societal well-being.

> ETHICS BY DESIGN.–
>
> An approach that integrates ethical requirements and recommendations from the design of NICTs.

> ETHICS BY EVOLUTION.–
>
> It is an approach that incorporates recommendations and ethical rules, in an evolutionary manner over time, throughout the lifecycle of NICTs, i.e. until its implementation and evolutionary use.

This book is intended to categorize ethical issues related to the digital environment, both from the point of view of the user and the designer of digital solutions and/or services. It invites reflection (what questions businesses can ask themselves about digital ethics) and suggests avenues for action. It is an approach that aims to provide guidelines to bring out the values that we want to collectively put forward to help legislators to formulate laws that will build a framework for AI. This repository is not exhaustive. It is intended to be general, open to all contributions and

7 This consists of integrating ethical rules and requirements from the design and learning of these NICTs, prohibiting direct or indirect damage to the fundamental values protected by the conventions.

evolving. It must be regularly updated to ensure its consistency and constant relevance as the digital environment and our technological knowledge evolves. It is intended as a reminder of the company's regulatory duties, which precisely define what is permitted or prohibited, and the sanctions that apply. The company has an obligation to comply, and this does not concern the area of ethics. However, the means by which it complies can be the subject of ethical reflection.

Finally, this book is addressed to all stakeholders involved in the development, deployment or use of AI, including organizations, companies, public services, researchers, individuals or other entities. This document should, therefore, be considered as the first building block of a discussion between these different actors toward an ethical, responsible, trustworthy AI aimed at protecting and serving in a beneficial way individuals and the common good for a better adoption at the global level.

1

Societal and Moral Questioning Around AI and Its Ecosystem

From autonomous cars to facial or voice recognition, artificial intelligence (AI) has developed and structured itself in a spectacular way over the last 5 years and is now part of our daily life and close environment. The widespread use of algorithmic applications feeds our imaginations, hypnotized by the promise of a better world, where the computing power of machines could reduce or even eliminate illnesses, accidents and crimes. At the same time, a growing doubt about AI is beginning to develop, portraying the technology and its exponential progression as a potential danger to the survival of humanity. Indeed, by the end of 2017, entrepreneur Elon Musk – the charismatic head of Tesla and Space X, among others – believed that efforts to make AI safe had a 5–10% chance of success. In doing so, he reaffirmed his earlier prediction that there was a risk that something very dangerous was going to happen within the next 5–10 years. At the heart of the concerns is technological singularity, a concept that predicts a runaway technological progress that could lead to the advent of a superhuman AI that would have autonomous capabilities to improve and evolve.

Moreover, there is no development of AI without the exploitation of gigantic volumes of data (Big Data). Indeed, we have to keep in mind that the computer without information can neither learn nor automate human action. AI stores the information that humans decide to give it. Failing to limit the collection of data, it would be appropriate to control its use and protection. As a result, the acquisition, storage, consumption and management of this Big Data are two decisive requirements for any contemporary society. So, with the digital revocation of AI and Big Data, all businesses and organizations have become aware of the potential that lies before them.

Now they want to highlight this relevant information and take full advantage of it. But how do you leverage accessible information while ensuring that you have a high-performance ecosystem in place to store, analyze and develop it? The emergence of algorithmic systems also raises the anxiety of a world guided and controlled by digital logic. To what extent can we leave the control of our contemporary societies to algorithms and those who design them? How can we guarantee the confidentiality of our private lives from the growing appetite of machines fed by the collection of personal data? How do we prepare for the upheavals and consequences that AI will bring about in many professional sectors? These are all questions that will be the subject of reflection in Chapter 2.

In order to pragmatically apprehend and understand these issues centered on human dignity and its fulfillment, we introduce four challenges and perspectives offered to society by AI (Floridi *et al.* 2018):

– whom can we become (autonomous self-realization): AI can assist in self-realization, i.e. the ability of individuals to develop in terms of their own characteristics, interests, potential abilities or skills, aspirations and projects;

– what we can do: AI allows us to improve and multiply the possibilities of human representation. Responsibility is paramount, given the kind of AI we develop, how we use it, and how much we share with all of its advantages and benefits. AI applications could help, if designed effectively, amplify and strengthen distributed and shared moral systems;

– what we can achieve (individual and societal abilities): if we rely on the use of AI-related technologies to increase our work capacities, we will be able to delegate certain tasks and especially decisions concerning autonomous systems that must remain at least in part subject to supervision and human choice. It is, therefore, becoming essential to find a balance between, on the one hand, the pursuit of the ambitious prospects and opportunities offered by AI to improve human life and what we can achieve, and, on the other hand, to ensure that we remain masters of these major developments and their impact on human society;

– how we can interact with each other and with the world (society, cohesion): AI can go a long way in dealing with such complex coordination, supporting more societal cohesion and collaboration, without undermining human dignity and without eroding human self-determination.

1.1. Use cases of AI

Now, AI applications are invading all sectors of activity and professional spheres of the company. Algorithmic uses are tirelessly multiplying and diversifying a little more every day. In each case, AI can be used to enhance human nature and its performance, creating actual opportunities that must be seized and well used (see Table 1.1).

Industry	AI use cases
Cities and local authorities	Increase user access to public services: – free up agent time by freeing them from repetitive tasks; – guarantee universal access to public service by breaking down the language barrier.
	Simplify citizens' lives and experiences: ensuring the efficiency of shared services.
	Optimize the management of the public budget: make the funding coincide with the actual consumption of goods.
Education	Better meet the needs of students: – prevent school and academic dropout; – support students outside of the institution; – support students in their choices.
	Transform learning: – promote learning to read; – propose personalized courses.
	Assist teachers: – allow teachers to spend as much time as possible with their students; – offer teachers feedback on their courses.
	Change the report to information.
	Streamline registration procedures.

Industry	AI use cases
Banking and insurance	In customer relationships: – develop commercial relationship; – save time; – speak the same language as the client.
	Reduce risk and fraud: fraud identification and anti-money laundering.
	Create new business models: – combine different businesses through data aggregation; – manage large numbers of investment simulations; – the creation of new products to support healthcare professionals in their daily practices.
Health	AI for better public health: detection and treatment: – revolutionize medical imaging; – harmonize care processes and support the doctor in their application; – generate alerts and reminders to healthcare professionals and/or patients; – review therapy and care planning; – recognize medical images and interpret them (radiology, ophthalmology, dermatology, etc.); – assist in paramedical care (paramedical humanoid robot); – assist in medical decision-making and establish predictive analyses via a diagnostic assistant; – enable communication interfaces between patients and healthcare professionals via a conversational agent (chat-bots) (conversational oncology); – monitor patients in real time and adjust their treatments to their individual situation; – make an earlier and more accurate diagnosis; – access to new knowledge; – improve the flow of the city hospital route; – reduce costs and pool resources.

Industry	AI use cases
Retail	Lead the customer to the store: – multiply access points: voice; – multiply access points: images.
	Transform the buying journey and improve customer relations: – make the customer's journey through the store more fluid; – provide an interactive customer experience; – adapt products to customers' needs and desires; – use facial recognition for various actions toward the customer; – offer customers the assistance of a personal digital assistant (PDA).
	Empower employees to do more by being more efficient: – facilitate the maintenance of in-store shelves; – alert employees when a customer needs them; – proposals targeted to the user. Optimize operations: – optimize the organization of stores and promote sales; – use customer data; – optimize inventory and inventory management; – optimize the delivery of orders and reduce costs.
Manufacturing industry	Optimize the production chain: – optimize the flow of production lines; – optimize the quality of the production lines.
	Improve the maintenance process: – predictive maintenance; – facilitate the work of maintenance agents.
	Strengthen employee safety.
	Make relations with suppliers more fluid.
	Obtain a better knowledge of the customer in order to better serve him or her.

Industry	AI use cases
Corporate Finance – CFO	Make sophisticated and reliable predictions: – obtain forecasts of the company's financial data and guide strategic decision-making; – determine employee bonuses.
	Free up human time (day/human): – answer questions from business teams; – manage the billing process; – check expense reports.
Marketing	AI for creative campaigns.
	Evaluate the performance of campaigns with a new level of accuracy.
	Promise a personalized experience.
	Constantly improve based on real-time feedback.
	A fluid client interface thanks to cognitive services.
Sales manager and customer relations	AI at the heart of the customer journey: – detect buying signals; – better manage the pace of customer interactions; – in store, detect abnormal behavior or optimize sales staff interventions; – automate information retrieval and call center operations.
	Gain time and efficiency in day-to-day work.
Human resources (HR)	Improve the recruitment process: – attract candidates through language analysis; – facilitate the application process; – identify relevant candidates.
	Giving employees the means to develop: – make interactions more fluid; – offer adapted and personalized training; – enable employees to find their next job with the company.
	Relieve HR managers of certain tasks: – contribute to risk and compliance analysis; – predict recruitment needs.

Table 1.1. *AI use cases by industry*

CHATBOTS.–

A software robot that can talk with an individual or consumer through an automated conversation service that can be carried out through decision trees or by an ability to process natural language.

Finally, we can envision three logics or ways in which AI will impact employment and the field of activity:

– a substitutive logic: it mainly concerns jobs that are not very highly qualified and implies an accompaniment toward new jobs for the collaborators and workers concerned;

– a rationalizing logic: this involves low-value-added jobs which, thanks to AI, could become less burdensome, for example, workers who work in assembly lines in factories;

– a capability logic: AI comes to enhance human capabilities by refocusing the activity on its added value and strengthening it (better performing and more relevant employees, distinctive skills that can be enhanced, etc.). Administrative and analytical tasks are then transferred to the autonomous algorithmic system to focus on human added value.

1.2. Digital environment

For the past 20 years or so, we have been living in a world where data are constantly multiplying, where visions and analyses have become infinite and where everything has become a sum of singularities and values whose structures seek to understand how to extract them. These large volumes of unpublished data are helping to store and build new knowledge, new perceptions and, therefore, new opportunities. We are now in an era of convergence between data, which can all become homogeneous, digitized and integrable, and with more correlation of senses. This is the digitization roll-out of the world, where databases and tools for storing and exploiting data on a large scale have been completely rethought and improved, considerably enhancing their operational performance. For more than a decade, online exchange platforms on the Internet have become the places where information, communication, knowledge and sociability converge. Digital culture represents the continuous path between the concrete and the conceptual, between the real and the virtual, adjusted by the evolutions of the digital environment. We are in the middle of an authoritarianism of immediacy, instantaneity and acceleration of the rhythm of life. This is illustrated by new Data Driven-centered approaches, where we visualize more than we model and where quantity takes precedence over quality. The consequence of this "datafication" is to provide the conditions and the means for governments and businesses to map society in a quantifiable and analyzable way, for an in-depth analysis of the reality that permeates people's daily lives, and even their thoughts.

> DATA DRIVEN.–
>
> An approach that involves making strategic decisions based on data analysis and interpretation. This approach allows for data to be examined and organized in order to better understand its consumers and customers. "Data driven" will, therefore, allow an organization to contextualize and/or personalize the message to its prospects and customers.

By its characteristics, AI irremediably leads to a considerable conceptual change around its digital ecosystem. Data warehouses are no longer at the center of the world. Many repositories and specialized tools support applications or new forms of analysis. Increasingly, data are coming from sources outside the infrastructure through application programming interface (APIs). As a result, the company that processes these data is more like a distributed supply chain.

> API.–
>
> In computer science, an API is a standardized set of classes, methods or functions that serves as a front end through which one piece of software provides services to other pieces of software. Software such as operating systems, database management systems, programming languages or application servers have an API.

Because of its systemic dimension, the digital revolution is causing major upheavals within companies and society. With the same disruptive force as the industrial revolution, it is transforming the business model, organization, corporate culture and strategy, and management style. As a result, companies using Big Data via AI are, therefore, faced with a new ecosystem that can be divided into nine segments (Kepeklian and Wibaux 2012, see Table 1.2).

Network of partners	Key activities	Offer	Client relations	Client segment
Host Datacenter HPC Manufacturer Cloud operator	Advice Calculation Storage	Collecting Processing logs Vertical applications Analyzing Visualizing Interpreting Structuring Storing	Self-service Training Support	Administration Media Industry Banking Health Distribution Etc.
	Key resources		**Distribution channels**	
	Size Performance Space		B2B A2B	
Cost structure			**Income flow**	
Platform maintenance Subscription Development Exploitation			Data valorization Sales/re-sales Batch/transactions IP, licenses, rentals	

Table 1.2. *Digital ecosystem of Big Data operated by AI*

CLOUD COMPUTING.–

Cloud computing is an infrastructure in which computing power and storage are managed by remote servers to which users connect via a secure Internet connection in order to deliver faster innovation, flexible resources and economies of scale. The desktop or laptop computer, cell phone, touch-tablet and other connected devices become access points for running applications or consulting data that are hosted on servers. The Cloud is also characterized by its flexibility, which enables vendors to automatically adapt storage capacity and computing power to user needs.

In addition, this digital environment is subject to many changes, both internal, particularly with:

– the explosion of dematerialized services and large volumes of data (mostly unstructured) coming in particular from the activity of Internet users and infrastructures;

– the decentralized ecosystem of companies and the heterogeneity of internal security levels justifying the implementation of data protection standards.

And of an external nature, in particular:

– the convergence of professional and domestic uses of digital applications. Indeed, social networks, discussion forums, Wi-Fi, blogs, instant messaging, Wikipedia, etc., are uses that are often incompatible with applications and needs within a company. This is why more and more professional infrastructures are establishing codes of good practice, rules of use and even ethical charters describing the value of these new information and communication technologies (NICTs) for the company;

– the complexity and amplification of the interrelationships with the environment that can lead to a diversification and dangerousness of threats. Therefore, it becomes fundamental that the structures rethink and evolve their data protection policy;

– the reinforcement of regulatory, contractual and legal constraints, relating to the need for increased transparency and reporting of Big Data. This situation incorporates a certain balance between increased disclosure of information, and control, on the one hand, of the authenticity and origin of the data, and, on the other hand, of the recipients of the information;

– the development of the extended enterprise leading to a transformation of relationships with customers, business models, suppliers, employees and partners. This context requires companies to broaden the scope of data protection beyond their natural and structural boundaries.

1.3. What is the place for human beings in this digital society?

NICTs are changing our lives and questioning the meaning of our lives in everyday life. Each technological innovation improves and optimizes practices, releases values and associated behaviors, and consequently inserts new social norms. NICTs force new considerations on the values and habits of our actions, because they give a greater number of individuals more means to communicate and interact with each other. Numerical criteria and their mathematical processing bring a new perspective of “digital humanity” by orienting people on the basis of self-referential normative indicators that are generally beyond our understanding. Thus, this new digital visualization does not describe our reality, but contributes to engendering it. The “natural” human being progressively gives way to the repaired, augmented, connected, instrumented human, etc., in which the term “human” is questioned in its very foundations. This is why, in their daily activities, connected citizens, connected devices, smartphones, online platforms and social networks that feed on digital data, are subject to diverse and multiple questions.

On the other hand, if we take the example of autonomous car driving algorithms, such as those used by the programmed option in an unavoidable collision situation, these algorithms will have to dictate decisions that may have a direct repercussion on the physical integrity, or even the life or death, of the people involved in the accident. This raises the question of who takes responsibility and who is accountable for safety and data control. Such a situation inevitably raises ethical questions about the criteria and parameters included in the "black box" of the algorithm that enabled the decision to be made. Indeed, citizens are all unaware of the algorithms they rely on in their daily lives to act and make decisions. Thus, they know almost nothing about the computer code, the data used and the choice of values that allowed digital giants to write these algorithms (Picard 2017). The opacity of the algorithm (Black Box Algorithmic Culture), therefore, generates a perception of low objectivity with intrinsically subjective decision-making involving implicit and/or explicit algorithmic biases. One can take the case of the modeling of the algorithm parameters, which leads by nature to a filtering, since it is a question of keeping only the parameters that seem "significant" to humans. As it is subjective parameterization, presumptions or judgments can be introduced into the code. Moreover, this bias can be silently propagated when a data set is used for a purpose other than that for which it was originally collected. For example, setting up an application analysis algorithm based on a candidate's postal code or experience can automatically exclude people who may also somehow be able to meet the conditions of employment.

A reflection on the design and the internal functioning of the algorithm processing is a natural choice. The question arises as to how their construction and use could be optimally regulated in order to allow the persons concerned to exercise their right of access to the data captured by these machines and/or objects.

Based on our knowledge of human evolution, at what level can we envy what our descendants will look like? At what horizon might we see significant changes? Until now, evolution has not been predicted; it has been observed. With the emergence of AI and soon of quantum computing, we will be able to have a predictive and, therefore, anticipatory approach to the natural evolution of living beings.

Digital technology, therefore, marks a break with the entire value chain and restructures our society and our daily lives, transforming them with a new culture and a new way of looking at the world. It is changing the way we live, consume, operate and work. It translates into a break with the past in the relationship and bond that each person has with others. From then on, these interactions make the system rethink each human activity. It is the beginning of a silent but very present revolution that is occurring right before our eyes. It is a new era of change and disruption where survival necessarily requires reactivity, adaptability, creativity and, therefore, innovation.

In the context of the generalization of open digital innovation and the transformation of the human condition, we are entitled to question ourselves from an anthropological point of view on the place of humans in this hyper-connected and digital ecosystem where new relationships to work and the transformation of work organization modes are emerging. Where is AI leading us? What place will the human being have in this hyper-connected and digitalized world where devices, robots, machines and other autonomous expert systems will interfere, respectively? Will AI leave a place for human beings, help them, make them dependent ("technological slavery") or make them disappear? On the contrary, does it threaten to make discernment, intuition and emotion disappear? To what extent can humans delegate their free will? Can citizens accept to be manipulated to transform their feelings, convictions or behaviors, and to be categorized or evaluated, without being informed? What impact will the development of AI have on our sense of ethics and the relationship between human beings and machines? Will humans see their status reduced to the state of sub-humans as some specialists claim? Will we move toward a form of "alter-humanity"? Is transhumanism inevitable for the survival of the human species? Or on the contrary, will this transhumanism not create a new hyper-connected human species, developing intellectual capacities that are currently useless and even pushing back the age of death? Insofar as it extends our time, our space and our senses, will AI disrupt our design of humanity and human emotions? How do we evaluate actions and creation from AI? How do we align the objectives of autonomous AI systems with those of humans? Would not the law be a brake on humanism? How do we prevent learning algorithms from acquiring morally objectionable biases?

Thus, the approach to ethical questioning is strongly associated with the conception of the place of human beings in their environment. Regardless of the products and services offered, digital technology has an impact on all business models and, as a result, on the ability to work and think together.

This necessarily implies that business leaders must think about the contributions of digital throughout the structure's value chain, in a transversal and systemic way, such as the analysis of cost structures, development and production processes, distribution channels, and pre-sales, sales and post-sales procedures. The complexity of this increasingly fast-paced world calls for new organizational models and a management style more oriented toward openness, autonomy, meaning and human values.

Without really paying attention, we are already surrounded daily by robots that take different forms such as a cash dispenser, an automatic ticketing machine to enter a parking lot, or even a cash register to pay for our purchases in a supermarket. These predictive or proactive algorithms are constantly invading our space on a daily basis. Obviously, this phenomenon is of particular concern in Western culture, which is much more fearful of this subject than in Eastern culture, such as Japan, which considers the robot as a real help and a benefit for humanity.

In fact, each industrial revolution has been accompanied by an upheaval in human habits and behaviors at the economic, political and social levels. This inevitably leads to a transformation of certain professions that have to be reinvented as well as the disappearance of certain professions that can be completely taken over by a machine. To this, we have to add the creation of new professions, leading to new markets, such as the qualitative selection of Big Data and the recycling of the latter. The phenomenon surrounding the "uberization" of society is a perfect example of this, with intelligent platforms that are moving at a fast pace because of algorithms that are taking over the work of intermediaries such as travel agency representatives, brokers, stockbrokers, cab drivers, etc., and soon, perhaps, pharmacists, opticians, doctors, judges, lawyers, notaries, etc. These intermediaries are usually located between the one who has the know-how and the client. Some professions will need to adapt to this new digital environment to evolve and meet the expectations of citizens and the requirements of business sector professionals. Thus, the ability to adapt, because of real-time information, is one of the keys to asserting oneself in this "fluid" society and deriving optimum benefit from it. From then on, it will be necessary to know how to adapt to the progressive reduction of salaried work, to the benefit of self-employed workers, coordinated among themselves. We can take the example of aeronautics, which is still, along with finance, one of the pioneering sectors in terms of innovation. Indeed, the 1980s marked the digital shift in aviation, with increased digitization in the piloting, management and maintenance of aircraft. Aviation professionals, such as pilots and air traffic controllers, saw their respective missions remain the same, but their functions and actions had to be significantly modified. This represents the consequences of an algorithmic neo-Darwinism that is spreading in all spheres of the planet, like a global pandemic.

AI is undeniably a factor of technological innovation, but it interacts like, until now, the human being has always been at the origin of progress, as much as intelligence has always been human and often collective. Social relations are increasingly deconstructed as they become more and more centered on people and interweaved in the dematerialized space. Added to this is the appearance of horizontal forms and local markets as modes of social and economic coordination. This trend is reflected in the emergence of new societal values and norms of behavior. Now, the digital world becomes a true vector of universal values.

Homo sapiens is evolving from a digital Darwinism to *Homo numericus* (or "digital human"), fully integrating all the signals, opportunities, obstacles and challenges of a ubiquitous digital environment, while being increasingly dependent on technology. Science fiction reveals that the future was at hand and that the digitalization of the human species was inexorable. In this context, human beings were thus led to "recreate" themselves, as transhumanism proposes. From then on, human beings become their own project, their own material on which to work in order to invent themselves, if not to (re)create themselves as they wish. This is why the "transhumanist revolution" leads to "post-humanism", i.e., the overcoming of humanity as we know it today, and thus, in the end, to the appearance of a new species. These new post-modern people are worried about "dead time" and unproductive monolithic action. They are constantly seeking to fill the void by a willingness to reduce their time of access to knowledge, by carrying out several tasks at the same time.

Some specialists, like the historian Yuval Noah Harari (2017), have even come to think that human beings will metamorphose into *Homo deus*, that is to say, the passage of humans into gods, because they envisage, thanks to NICTs, becoming immortal (by defeating death), finding the magic formula for happiness and increasing their intellectual, cognitive and physical capacities (by modifying their brain and body). Indeed, we can imagine that with the arrival of the next industrial revolution centered around quantum applications like quantum computing, we will be able to, in the medium term, "download" human consciousness into a computer. This presupposes the existence of a mysterious immaterial substance (perhaps quantum information?) capable of freeing itself from its material support (the human body) to an almost ethereal environment. This will correspond to the final stage of our *Homo deus* for which we have reinvented the soul: namely, an atheist soul whose only God is technology and whose digital ecosystem will play the role of paradise or hell. This ultimate stage of transformation would allow the *Homo deus* to glimpse his or her own finitude in order to reach the end of his history. In an interview with the BBC on December 2, 2014, the British astrophysicist Stephen Hawking even declared, "The development of a complete AI could put an end to the human race".

Finally, this computer reductionist approach to the world contributes to a break between consciousness and intelligence, with algorithmic devices and systems that, without being aware of their existence, are becoming increasingly self-nominal and intelligent. In fact, we can take the example of the AI AlphaGo, which by beating the world Go game champion had not celebrated its victory. It did not even know that it was playing and, thus, become aware of what it had achieved. Therefore, is not the "datafication" of our world likely to change the order of priorities and values of humans between intelligence and consciousness?

Moreover, the relationship of the human being with time and space tends to tend toward instantaneity and ubiquity. This can be illustrated at the level of the modern company, which feels the need for real-time control in order to be able to constantly adapt to unstable economic and social environments and volatile customers. This Business Intelligence (BI) will have to take full account of unstructured data or even Big Data. This will therefore require the development of new methods for managing and studying data, as well as new tools, operating methods and know-how.

BUSINESS INTELLIGENCE.–

BI is a technology process that analyzes data to present information that can be used by executives, business owners and other users to make better decisions. It encompasses the various tools, applications and methodologies that enable the enterprise to collect data from internal systems and external sources, prepare it for analysis, develop queries and apply them to that data. The objectives and benefits of BI are to accelerate and improve decision-making, optimize cross-enterprise business processes, increase operational efficiency, generate new revenue and gain competitive advantage. Finally, these programs also help to identify market trends and business issues that need to be addressed.

As a result, the traditional value chains of an organization have become more complex and inevitably require a more matrix, systemic, cooperative, and transversal management and steering. As a result, one of the major human challenges ahead lies in the ability and capacity of each individual to evolve and rethink and structure oneself in order not to be left behind by this digital (r)evolution. Such a challenge will inevitably require a reinforced involvement around initial and continuing education, the reform of universities, orientation, and more generally learning. In addition to the development of skills adapted to the transformations of society, the digital revolution is leading to a more global rethinking of pedagogical learning methods. Technologies provide the opportunity to build new devices, for example with digital universities by improving teacher training, to rethink the production of digital resources and to promote the development of distance learning. On the other hand, there is a tendency to say that whoever holds the information holds the power. However, in a contemporary "information society" inhabited by people in constant search of information and knowledge, we can see that human beings' power of influence is increasing both economically and politically as a result of NICTs.

In our opinion, it is very likely that the constant development of AI will incite people to want to enter transhumanism to become an "augmented human" who can still compete with robots and intelligent machines. This fear of AI should result in the acceptance of each individual to integrate a brain augmentation via technology. To illustrate this vision, we can quote the words of Elon Musk (founder of Space X, Tesla, PayPal, etc.), in June 2016, who said that it was urgent for humanity to hybridize its brain with technological components to avoid any vassalization by AI. The integration of digital giants will be increasingly important and fast, due to a particularly efficient platform model:

– based on the logic of the economy of sharing with a direct relationship between supply and demand;

– facilitating social and/or commercial interactions through the use of Big Data on user behavior.

In the reflection that drives us, it seems indispensable to focus our discussion on the notion of digital culture, which must constitute one of the cornerstones of tomorrow's society. Indeed, as a social construction, culture influences people, their behavior, their ability to act and even react, and their habits. Our ability to acquire and study a large volume of data is evolving. We are now able to discern reality "at scale", in all its complexity and dynamism. In some ways, this shift resembles the movement of Newtonians into the Einsteinian world of gravity, that is to say, a transformation of the way we perceive the world around us and the meaning we give to it.

Our cultural references are different and experience is essential in order to understand and act effectively on the ground in countries, societies and civilizations where references, traditions, sociopolitical frameworks, customs, and traditions and relationships differ from ours. This cultural knowledge is indispensable as a reference point for the decisions that will be taken to respond to a given problem. It is, therefore, advisable to carry out an ethical reflection on NICTs. Consequently, it is fundamental to take these cultural differences into account so that the uses remain consistent with our conception and understanding of society and its citizens. It is, therefore, essential that we arm ourselves with a critical culture of numerical computation in order to play with them, to divert them, to take advantage of their relevance and their predictive power, in particular, without letting ourselves be impressed by the worlds they seek to impose them on us. The development of this culture is fostered by encounters, by individuals who wish to share their skills, know-how, and life skills. Every citizen has a responsibility to take and assume regarding the content of the future of society, with the search for a balance between good and evil always being the central issue. This is why it is fundamental to make citizens aware of the challenges of the digital world from an early age, by developing vigilance and in-depth education around digital technology in order to really grasp its opportunities and challenges as well as its risks and deviations. We can see that risks can be the source of several factors at the same time: human prejudices, technical risks (for example, through the lack of training, verification, and validation of the functioning of the algorithm), risks of use linked to the inadequate implementation of the algorithm (for example, an accident in the street that takes place because the users have not understood that it is not an autonomous car but an assisted driver, or in the case of an incorrect understanding of a result shown by a decision support algorithm), or in the face of internal and external dangers on the manipulation of the algorithm in the face of its vulnerabilities (for example, in the case of biometric authentication).

This cannot be achieved without the development of bridges between research, education/training (universities, engineering schools, business schools, etc.) and the business world. This can take the form of concrete actions such as:

– rethinking digital learning and orientation in school;

– inclusion in civic education courses of an awareness of the societal and ethical issues and risks surrounding NICTs;

– encouraging engineering, business, and political science schools and universities to develop multidisciplinary training (initial and/or continuing) adapted to the digital professions.

In addition, this revolution around a new world impacts all aspects of life, society, and all areas of our business: e-business, energy, health, urbanization, politics, ecology, citizenship, etc. This digital environment is then conducive to the development of new players like data scientists – sometimes called quants (short for quantitative analyst) – who perform quantitative analyses. It should be noted that we sometimes speak of "algorithmists" to identify people who are quants and who have integrated a specialized profession and act as independent audits of Big Data studies. The link between data scientists and "algorithmists" is somewhat similar to the link between a person who has studied medicine and a person who is a general practitioner or surgeon.

Finally, man is in the process of outsourcing (outsourcer) a human capacity via NICTs in order to carry out other actions. We can take the example of cell phone numbers, which are now directly accessible in digitized directories, and which we no longer bother to memorize. Is this going to change the profound nature of the human being, especially with the arrival of transhumanism, i.e. the augmented or even improved human being? Will it lead to the appearance of other human beings totally connected with the smallest particle of their environment, which is itself connected in a network? This brings us back to the theories of singularity, the concept of which indicates that, from a hypothetical point in its technological evolution, human civilization will be subject to technological growth of a higher order. Data are a kind of atom in the digital world that people will have to transform and manipulate in order to evolve. Since human beings are themselves sources of information, it is up to their survival to better understand the full potential of this hyperconnectivity. Based on this observation, it is up to us to reflect on the future status and positioning that can be attributed to us in a neo-Darwinian environment where virtual and real worlds will merge together. It is still too early to say that AI will never replace human intuition and discernment, but it is up to us to define today the path to take if we do not want it to become a threat to human beings. NICTs therefore announce more than a transformation, and it is a real revolution in the field of our individual liberties that impacts the future of human nature as we know it today.

1.4. Technological and societal issues

Difficult to grasp and control, algorithms optimize and guide our choices and decisions. But who knows the content and functioning of these algorithms? How do these new algorithmic powers transform the professional practices of society? How does an algorithm choose its actions and how does this choice compare to that of a human being? Will AI have the autonomy and human creative individuality to replace it one day?

Take a case that is often cited, that of replacing obsolete, or even useless, doctors with intelligent machines. At present, the answer would tend to be "no", because health and illness are strongly influenced by subjective, emotional, and social parameters and criteria to which conventional machines have little access in terms of skills. Indeed, disease is an ill-defined problem. Technological knowledge cannot define and totally represent the event of a single patient's illness. A deliberative doctor–patient relationship illustrated by associative and lateral thinking is decisive for healing, especially in complex cases and when the risk of undesirable effects is high, because each patient's preferences are different and varied. Thus, the use of AI, based on the belief that symptoms are measurable, reaches its limits when confronted with social, emotional and non-quantifiable variables related to the disease. Questions such as "Why now? Why me?" and "What did I do wrong?" are essential for the patient, especially those with cancer. Today's AI is increasingly capable of performing tasks and procedures that were once the prerogative of traditional practitioners, including prognosis, diagnosis and treatment. Although they increase the capabilities and knowledge of physicians, so-called “weak” AI cannot fully replace them. Machines cannot take into consideration the human desire to always associate disease with the task of living a life, which is linked to the human context and to subjective and non-measurable indicators of disease. Even today, health professionals are still better able to take care of the patient as a whole person, as this requires an increased knowledge of social and psychological relationships and normativity. The health professional can build relationships with the patient as a human being and acquire a comprehensive knowledge of the individual’s illness in relation to his or her life. Such knowledge requires ideals such as respect for human dignity, trust, responsibility, courage and empathy, which are not easily accessible to today’s intelligent machines. At present, there are no algorithmic systems yet adapted to this type of comprehensive care based on emotions, non-verbal communication, human values, moral principles, personal preferences, prevailing social circumstances, etc. It is clear that today’s machines are not capable of taking care of patients alone, in the sense that they do not show dedication and moral support to a person. Some sophisticated robots can reveal empathy as a matter of form, just as human beings can behave well in social situations while remaining emotionally disengaged because they only play a social role. We strongly believe that with the arrival and development of the so-called “strong” AI that is at the same time evolutionary, self-learning and endowed with emotions and artificial consciousness, the factors that are not yet apprehended by machines will be in a few decades.

Moreover, will technology set the standard without any democratic debate? What kind of ethics are needed to make digital technology a major element of our democracies? Are we prepared to build, or even develop, ethical-digital links, to take into account the interrelationships and interdependencies of our main societal and moral problems?

These successive and complex questions are the place where the respect of human values and the binding limits of concrete decisions to be taken are constantly encountered. Numerous questions relating to the large volumes of data, their construction, collection, access, storage, exploitation and use, particularly by information technology (IT) devices, arise in all sectors of activity.

Therefore, we are entitled to ask ourselves whether this digitization of society has its own ethical problems. The issue of the individual’s private life is central. In fact, the cross-referencing of databases from various sources, which probably does not guarantee the same level of security, can lead to questions about the level of risk created for the source person. In other words, does the crossing of databases and the increase in the amount of information lead to an increased risk of invading privacy and lead to an illicit use of the data? Do privacy and confidentiality have the same value on the Internet as in everyday life? This question is all the more important since the vast majority of business models around the digital world rely, to a greater or lesser extent, on some form of data monetization.

On the other hand, we notice that at first glance the property of a digital data is by nature quite distant from the property of a physical object in the real world. Indeed, in everyday life, the owner of a car has a precise title of ownership and has access to his "object" when he wishes, whereas in the virtual world, this situation is much more complicated since users themselves do not store their virtual objects, in particular their digital data, in a privately administered database, and until now there is no attestation certifying that a user is the originator of his or her data. In fact, the data are scattered in data centers belonging to the digital giants (Zouorhi 2017). So, what are the concrete and realistic means available today to the user of the data to prevent it from being used, studied, or even sold by a third party? Currently, probably none!

ADMINISTERED DATABASE.–

The administered database is supervised by an administrator who reliably organizes and manages the company's data management systems. The administrator must ensure the consistency, quality, security and permanent accessibility of information. He/she participates in the choice of software packages and the implementation of the company's databases. Then he or she who installs, configures, administers and optimizes the database(s). Beyond the technical aspect, he or she takes into account the entire environment of the company as well as the needs and requests of the users. In the absence of administration, the impressive volume of unstructured data generated annually within a company can be costly in terms of storage. At the same time, unmanaged data can also pose a liability issue, for example, if the information cannot be located as part of a compliance audit or legal action.

What is the qualification of the data? Should data be considered as immaterial goods that are subject to appropriation (property right)? Should data be considered as an extension of the person who cannot be the object of a patrimonialization or, on the contrary, as a right of the person (personal rule)? But the main question is who owns the data. Are they the property of the end user or of the person at the source? Does the data become the property of the company collecting the data? Thus, we are led to wonder who will be the rightful owner and who will benefit from the associated wealth? Those who are at the origin and must collect the value? Those who store it, exploit it and infuse it with value?

The various market players have not yet all agreed on this subject, depending on the country. Indeed, it is useful to recall that unlike our Atlantic neighbors who are subject to Common Law, in Europe and therefore in France, no property right has been recognized on the data, and therefore no right of purchase or sale. Under these conditions, according to the European approach, the data embody an emanation of the legal personality of a natural person. It belongs to the family of subjective rights[1]. Like the identity of the individual, it benefits from the same legal protection as information associated with the private life of individuals (Williatte 2017).

The central question here is, therefore, knowing what means we can give to the true "owner" of the data so that he or she can really win it back. In order to try to answer this question and perhaps provide a possible solution, it seems fundamental to us to explain what we really mean by "owner". The very essence of ownership is the ability to have access to a tangible or intangible good, and to have full powers of use of it. With the emergence of blockchain technology, it seems, today, possible to reach a degree of sovereignty for the user. One example is the uPort "self-sovereign identity" device, which enables identity management by drawing up attestations on the identity of an individual or data as well as on claims that make it possible, for example, to prove that an individual and data are associated (Zouorhi 2017). Under these conditions, the blockchain lets us envision a world where we could remain the owner of our data while making it available to others, a kind of virtual usufruct. The user who becomes the real owner could generate data and decide who will have access to it.

---

1 The expression "subjective right" refers to all the prerogatives, advantages, or particular powers that a subject of law, whether a natural or legal person, enjoys and can avail oneself of.

In the end, it might be preferable, even essential, to consider digital data as an intangible asset and to designate the Internet user or source user whose data are associated with it as the true owner of these data. This is all the more topical with the development of the blockchain tool, which will help give users the means to reclaim their data.

On the other hand, questioning the construction, functioning and purposes of algorithms through an ethical prism is not binding and can even quickly lead to added value, meaning, transparency and trust among citizens. As a result, where uses are unpredictable, discussions about NICTs open up perspectives for possible uses, and would, therefore, help to prevent certain abuses.

In this context, this "datafication" of our society leads to a whole series of questions, especially around Big Data, algorithms and AI such as:

– What are the changes? What does the exploitation and the analysis of Big Data bring within a structure?

– What place do we want to give to data in our lives?

– How does "massive data" impact daily practice? What changes can be expected in client services?

– How are big data used by companies?

– How do we associate the different data sources? What is their attributed value? What are the different transformations performed? What are the intermediate data produced?

– How do the results of big data project analyses guide the company in its daily activities?

– Do structures that use personal digital data resell the latter to external third parties?

– Where can reliable digital data be found and how can they be used?

– Should judgments and decisions be guided only by a self-learning expert system, AI, or algorithmic processing based on correlations? Are we ready to blindly undergo these NICTs or do we prefer to wait patiently for their final emergence?

– How do we reconcile transparency and intellectual property around these algorithms?

– Can Big Data represent "objective truth" or is any interpretation necessarily biased by some form of subjective selection and filtering, or by the way the data is "cleaned" (Bollier 2010)?

– Is an algorithm more reliable than the human decision, and should it substitute and take precedence over the human decision? Are algorithms the new decision-makers?

– How do we build one's free will in a world governed in part by algorithms?

– Should our habits and gestures that characterize and individualize us be dependent on values and reference data common to all of us that come from algorithms?

– Should people be informed of the risk of disease following genome sequencing?

– Do algorithms lock us into an informational cage, jeopardizing cultural openness and democratic pluralism?

– All this accumulated Big Data has value; how and where can we find this value to improve a business?

– Is it necessary to adapt a company's organization to these technological evolutions?

– Is the algorithmic solution still of acceptable quality if the instance is disturbed (robustness), if the system is dynamic (fault resistance), etc.?

– Have professional codes of ethics taken into consideration the integration of NICTs?

– Does the introduction of algorithms in the field of health or insurance impact the doctor–patient or insurer–insured relationship?

– Are not algorithms for the exploitation of digital personal data going to lead to a new wave of inequalities, categorization, discrimination[2] and injustices within society?

---

2 Particular attention must be paid both at the design and usage stages, especially when such processing is based directly or indirectly on "sensitive" data. Such data may be considered as such, for example, alleged racial or ethnic origin, political opinions, socioeconomic conditions, religious or philosophical convictions, trade union membership, as well as biometric data, genetic data, data concerning health or sex life, or sexual orientation.

– Is an algorithm loyal to society? Does its purpose serve a collective interest and protection of individuals or a more personal interest? Is it trustworthy?

– Does the algorithm respect the human dimension of the data it exploits?

– To what extent must professional practices around digital technology meet the criteria of objectivity, neutrality, and/or rationality?

– How far is it possible to value a company's approach and ethical commitment in the use of its Big Data?

– How can the need for sharing and personal privacy be combined when storing digital data in a cloud computing environment?

– How can the collective interest and individual protection of individuals be combined?

– How can an ethical culture be transmitted to digital professionals with an awareness of the latter?

– Is ethics inherent to the emergence of AI?

- Who owns the knowledge that emanates from expert systems (like Watson): the owner of the data, the structure that exploits this information, the publisher (like IBM), the citizen?

- How can digital transformation and ethics within a company be reconciled?

- Is there a recognized reference on good ethical practices around the building, operation and use of algorithms?

- Will NICTs lead to a new wave of social and technological inequalities and injustices?

- Are the results from algorithms scientifically relevant and reliable?

- How can new forms of analytics and applications to exploit both new and old data be enabled?

- Are the new digital players (Big Tech, operators, etc.) going to interface with specialists in sectors such as health, insurance, finance or aerospace? Will there be more competition or partnerships?

- The elaboration of data can be a relatively complex process (simultaneously combining human activity, sensors and one or more processes); which contributors are likely to claim a property to the associated value by the exploitation of the data?

- What are the responsibilities of each of the (public and/or private) actors involved at a given point in the data life cycle? Who is responsible for the algorithm and the exploitation of digital data?

- Does the NICT user have real control over his or her data? Is he or she aware of the possible resale of his or her data?

- How can the structures be regulated when everyone is able to launch their own online excavations and data extractions using a specific algorithm?

- Which public and/or private bodies must guarantee optimal efficiency and security while ensuring the democratic and transparent nature of digital data processing?

- How will the economic and financial model of a 4.0 company evolve?

- How can we produce wealth through digital technology, while creating an economy of trust, which is essential for new uses to develop?

> BIG TECH.–
>
> The largest and most dominant companies in the IT industry.

In addition, the question about the quality and intrinsic nature of the data also contributes to reinforcing this apprehension and concern of society. Is the integrity of the data produced and received guaranteed, in particular by an electronic signature device?

Digital players have found that data mining can increase the number of errors if the data are not integrated into the database. Numerous malfunctions could then taint the veracity of the data and the processing, if the data came from the wrong sources, if it was duplicated, or if it was already obsolete or out of date. This phenomenon is further reinforced by the fact that Big Data is increasingly coming from sources outside the company that uses it. To combat this situation, publishers are implementing new solutions that improve the visualization and traceability of the source of the data and reduce the error rate. It is also observed that new players specializing in the cleaning, analysis and selection of external data have emerged.

Thus, we can list issues of a technical nature (interpretation/ understanding of algorithm decisions and their impacts, respect for privacy, data protection and free and informed consent, security (including cybersecurity), management of voluntary or involuntary biases, and of a socioeconomic nature (with the accentuation of imbalances in the distribution of wealth, pressure on employment and the transformation of trades and professions).

FREE AND INFORMED CONSENT.–

Free consent means without coercion or threat. It means giving informed consent, and also without impairing your faculties. For its part, informed consent means that the person has received all relevant information about what is being proposed so that he or she is aware, as much as possible, of the different options available to him or her and the risks and consequences associated with each option. It is an informed decision. For example, in the health field, this means that when a doctor proposes care or treatment to her patient, she has an obligation to inform the patient about the nature or purpose of the proposed treatment or hospitalization, the possible risks and side effects, the consequences of a general refusal of care and, thus, can force you to leave the institution.

CYBERSECURITY.–

Cybersecurity, which concerns the digital security and sovereignty of each nation state, presents economic, strategic and political challenges that go far beyond the security of information systems alone. It also concerns management information systems, industrial information systems, embedded computing and connected devices. Cybersecurity must be approached in a holistic way, taking into account economic, social, educational, legal, technical, diplomatic, military and intelligence aspects.

Finally, faced with the emergence of robots, automatons, or humanoid machines that would look more and more like humans, new risks for the future society are appearing, such as:

– the risk of a certain servitude and alienation of human decisions, imposed by the algorithmic system that would decide for us what is right or wrong, leading to a totalitarian takeover of our daily life and more globally of our life;

– the risk of a society that is increasingly resigned and lazy because of too much digital dependency would push individuals to choose the dictated solution without going any further;

– the emotional risk, due to rejection or too close a relationship with the robot.

Under these conditions, the separation between the people who understand these machines and can thus maintain their autonomy and free will without being subjected to the choices of AI and the others would be more and more consequent. Thus, it becomes essential to pay special attention in the human–machine relationship by integrating ethical rules around the learning of the machine and its consequences, education to AI and its operation, good behavior in computer programming, verification of compliance with these rules by the robot, etc.

1.5. Ethical and moral issues

This digital revolution – illustrated in particular by the growing emergence of decisions made by algorithms contained in platforms, machines or technological objects – raises a whole series of ethical questions centered on the protection of individuals against the disclosure and exploitation of their personal data, and on a possible "over-mathematization" of society. Indeed, the consumer often loses all control involving the dissemination of his or her data. Who uses the data? For what purpose? What is the free will of data users? Where is it stored? How can the individual maintain control over his or her personal data? How can we guarantee that the data that we issue or that transit through us and the NICTs cannot allow formal identification or interpretation that escapes us? What is the degree of transparency? Open source of the code and documentation? The system could be explicable in a clear and popularized language in order to describe the way it produces its results, for example, by communicating on the nature of the services offered, the tools developed, the performances and the risks of error?

OPEN SOURCE.–

The designation open source, or "open source code", applies to software that is licensed according to criteria specifically established by the Open Source Initiative, i.e. the possibilities for free redistribution, access to the source code and the creation of derivative works. This source code is made available to the general public and is generally the result of collaboration between programmers.

With respect to digital data, the ethical framework for people and structures can be summarized in four specific concepts (Davis and Patterson 2012):

– Identity: what is the link between our offline identity and our online identity?

– Confidentiality: who should control access to data?

– Ownership: who is the true owner of the data (subjective right, transfer of ownership, etc.), who has the rights to transfer them and what are the obligations of the individuals and/or structures that generate and use these data? Does our existence consist of innovative acts on which we have copyright or other design rights?

– E-reputation: how can we identify which data are trustworthy and reliable?

The primary challenge is to give citizens back control of their information. At the heart of this reflection lies the positioning of the cursor around the responsibility and ownership of the data by the actors. Indeed, we are led to wonder about the responsibility of the individual in the case where the anonymization of data leads to a certain invisibility of the individual who can clear him of certain rules of decorum.

ANONYMIZATION.–

The anonymization of data (*a fortiori* personal data) consists of modifying the content or structure of these data in order to make it very difficult or impossible to "re-identify" the (natural or legal) persons or entities concerned (which implies clearly defining what the concept of identifiability means in this context). Anonymization can be carried out "at source" by the entity producing the data, or "at output", after processing and analysis.

Finally, the fact that AI will have a major impact on society is no longer questioned. Now, the current debate is more toward to what extent this impact will be positive or negative, for which person, in which way, in which places, and on which time scale? In other words, it is no longer time to wonder if AI will have an impact – the answer is inevitably yes – but rather to know who, how, where and when these positive or negative consequences will be felt (Floridi *et al.* 2018).

Finally, one of the central questions is that as a citizen we can choose in which society we want to live in tomorrow, and what will be the services emanating from the digital ecosystem that we want for the future of our society? How can we control and regulate algorithms and their uses tomorrow? Who will ensure that the defined digital rules will be well respected? Is there a recognized repository on the good ethical practices of these processing algorithms?

On the other hand, if "the code is ethics", this means that the people who integrate ethics in NICTs are the editor, the designer and the developer (or the implementer) of AI. Therefore, one may wonder if it is ethically acceptable that only the manufacturer decides the parameters to set up to obtain a given answer? Which organization, which roles, which rules, which policies around and on algorithms? Should not this ethical character of politics be given back to independent multidisciplinary committees, or to public or private regulatory bodies in charge of studying the relationships between human rights, technology, markets and economic models in the 21st century?

These are all questions that deserve to be examined and debated with urgent and profound attention so that these decisive issues around NICTs, as well as concerns relating to the future shape of our society, are the subject of enlightened and enlightening choices. Consequently, it seems essential and fruitful to us to approach these questions in an ethical light in order to nurture a rich, open, constructive and evolving reflection. Indeed, ensuring socially preferable outcomes from AI depends on resolving the conflict between incorporating the benefits and mitigating the potential drawbacks of AI, in short, simultaneously avoiding the misuse and underuse of these technologies. Compliance with the law is simply necessary (the contracts that are required), but significantly insufficient (not the maximum that can be done). By analogy, it is the difference between playing by the rules and playing well, so that you can win the game. Taking an ethical approach to AI gives what we define here a "double benefit". On the one hand, ethics enable organizations to leverage the social value that AI enables. This is the advantage of being able to identify and exploit new opportunities that are socially acceptable or preferable. On the other hand, ethics allow organizations to anticipate and avoid, or at least minimize, costly mistakes. This is the benefit of preventing and mitigating actions that prove to be socially unacceptable and, therefore, rejected, even when they are legally unchallengeable. It also reduces the opportunity costs of choices not made or options not entered for fear of mistakes.

The double advantage of ethics can only work in an environment of public trust and accountability. Public acceptance and adoption of AI technologies will only be possible if the benefits are seen as significant and the risks as potential, but avoidable, minimizable, or at least something that one can be protected from, through risk management (e.g. insurance) or redress. These attitudes will in turn depend on public participation in the development of AI technologies, openness to their mode of operation, understandable, widely accessible, regulatory and repair mechanisms. In this way, an ethical approach to AI can be seen as an early warning system against risks that can endanger entire organizations. The clear value to any organization of the double advantage of an ethical approach to AI amply justifies the expense of commitment, openness and contestability that such an approach requires.

The ethical questions relating to AI that come up most often concern autonomy, decision-making capacities, learning, the acceptable level of delegation to AI, the preservation of collective interests, emotional and social interaction, imitation of the living, repair and augmentation of the human, the dilution of responsibilities (the designer? the manufacturer? the vendor? the operator? the user? the autonomous agent? etc.), the creation of a "legal personality" for an autonomous digital agent?

Today, society has not really established any one-way rules or guidelines to help incorporate the moral standards or human values that humankind brings to AI. With the exponential growth of digital technology, it is now urgent to find a consensus around a universal ethical foundation in order to be able to communicate and explain – for the sake of transparency – the actions of autonomous intelligent systems. The challenge is to acquire an optimal degree of confidence, given the scenarios in which human beings use them.

In this context, those responsible for AI applications must integrate, right from the design stage, the challenges and risks they pose to civil society. In order to respond to this, we have identified two non-exclusive approaches that can be differentiated:

– ethical or moral AI: this approach models and programs elements from ethics in an intelligent application, thinking about how to allocate the machines of ethical principles and rules or a process of solving ethical problems they may encounter in order to become ethically responsible on the basis of a proper ethical decision system;

– trustworthy AI: this approach considers ethics less as the integration of ethical reflections within the code, which is endowed with general properties and concepts allowing to control the conformity of automatic decisions with the values agreed upon by the social group that uses them.

In this book, we are more in favor of the first approach, which seems to us much more promising, evolutionary and sustainable over time. It fits perfectly into our approach of Ethics by Design and then Ethics by Evolution that we want to implement in AI.

Indeed, because of the conceptual complexities relating to moral "values" and "principles", it is difficult to envisage and apprehend calculation structures that directly correspond to universal human values. However, if we associate ethical norms and rules to these universal principles, we can more easily imagine integrating explicit ethical norms in AI. Indeed, these ethical rules can be identified as pragmatic instructions to act in a defined way in defined contexts. One can then speak of ethical decoding and encoding of AI. The design of AI techniques must imperatively embody ethics. According to Asimov's vision, it is essential to integrate ethics in technology, i.e. the upstream consideration, by anticipation, of ethical dilemmas and issues that could go against the expected innovation trajectories. This means first of all, for users, to obtain a consensus on common ethical principles, then to encode them in AI applications, in order to ensure that the mediations they carry out respect them.

Once ethical principles and rules are recognized, they will need to be incorporated into AI technology so that their behaviors are consistent with them. More generally, NICTs must also satisfy the values on which societies rely on them. These examples of AI are thus subject to increased vigilance because of major specificities, decision-making autonomy, and learning that impact three fundamental social issues: transparency, security and responsibility. The challenge is to design algorithms capable of functioning in situations involving ethical considerations. This generates two risk categories: risks associated with design flaws and risks associated with learning.

Thus, making sense of AI highlights three angles of view:

– the *guidance* provides AI policy and strategy;

– the *meaning* reveals the societal dimension of AI, which should not be an end, but a tool at the service of society leading to the complementarity between human beings and intelligent systems, the explicability of algorithmic systems and the development of an inclusive AI;

– the *explanations* concern a collective reflection on the objectives pursued and their merits.

The establishment of such a system of ethical standards is not simple to put in place, and raises some obligatory questions (Floridi *et al.* 2018; see Appendix 4).

Under these conditions, we recommend:

– first, an inclusive approach of the stakeholders from the elaboration of AI (design) in order to bring transparent signals (capacity of inspection and/or explanation) on the nature of their purpose and their behavior toward the actors of the society: this is Ethics by Design;

– in a second step, to widen this approach throughout the development and lifecycle of AI (implementation and use), by making ethical criteria and metrics evolve as the machine learns and evolves. This is what we call Ethics by Evolution.

Thus, this method of proactive ethical inclusion of users and their interaction with AI will increase, over time, the overall confidence and reliability of algorithmic systems.

2

The Ethical Approach to AI

From the mastery of fire to that of the atom, from printing to the steam engine, every technological advance of humanity has brought both hope and anxiety. It is noticeable that gaps (discrimination, excess of power, inequalities, etc.) in the NICTs that worry us so much are only the expression and translation of the drifts that have been perpetuated since the dawn of time in human civilizations. It can be said that technological innovations have the moral value that we attribute to them by the use we make of them. Each industrial revolution leads to a humanist revolution that generally marks a turning point in civilization and a reworking of the societal model. If the socioeconomic system no longer meets humanity's requirements for survival in this increasingly digital environment, then we must reinvent and build a social project based on another system. This fourth industrial revolution – based on Big Data and intelligent machines – is not at all like its predecessors, with incomparable breakthrough power. We are in the midst of a moment of uncertainty, confusion and changes, where humanity can no longer project itself into the future, as when the sea and the sky merge so that we can no longer see the horizon line. For the first time in the history of mankind, it is the future of humanity as we know it that is being played out through the lines of code of intelligent machines. The pace of technological innovation and the temporality of its worldwide launch, supported by the digital economy, is clearly outpacing the speed of human awareness. In these conditions, it is high time to conceive and structure digital ethics and in particular one related to AI.

2.1. Definition of ethics

The word "ethics" originates from the Greek word *ethos*, which means "morals" (Cicero) and "habits" (Plato and Aristotle). Ethics concerns the "habitat" and the "character of a person". Thus, the way we inhabit the world represents the way we are somebody. The expression "being inhabited" takes on its full meaning and symbolic value. Seen from this angle, ethics is a reflection on the habits that must be contracted in order to make a space inhabitable. Ethics thus brings a questioning on the values that underlie action, conducive to a conflict of values in a world of ideas. It "naturally finds its source of reflection in action" (Hervé 1997). Its objective is, therefore, to give meaning to actions. Ethics is an individual disposition to act according to the virtues in a given situation in order to seek the right decision. It only makes sense in its own situation in which it admits argumentation, discussion and paradoxes.

Ethics refers to the conditions of a good life, for oneself and for others. It is "the desire for a fulfilled life, with and for others; in just institutions" (Ricœur 1991). It is the order of interpretation and/or practice. The ethical act is first of all a response (from the Latin *respondere*: to respond from/to, hence responsibility) to a limited and complex situation. Ethics has three main functions, namely the determination of what morality is, the knowledge of the reasons justifying a person's effort to live morally and the application to practical life of the results obtained in the first two functions. One of the essential characteristics of an ethical decision is that it is a tailor-made decision, a compromise resulting from the consideration of all the factors that determine a particular situation. An ethical decision is, therefore, always based in part on fundamental rights, the content of which is also adapted to each particular situation they are intended to govern. On the basis of this observation, it is possible to consider ethics as a vector for the guarantee of fundamental rights in a given situation.

Every person seeks the values that govern him or her, chooses the principles of action that should prevail, watches over the conditions of their implementation and makes himself or herself sensitive to their reality. Ethics is above all an adventure, a compass, the search for an interpretation and a posture, adequate to the reality to which one belongs. We confront reality through the prism of our feelings, emotions, objectives, thought patterns and representations, which challenge and mobilize us. Interpretation and analysis involve both intellect and affect. It is the combination of these two components that gives value to reality and is articulated with ideas or processes of ideas in which it finds a coherent meaning. It is this complex system of valorization and devaluation, between rational and sensitive, that it is important for us to apprehend and appropriate. This necessarily involves a system of mediation that plays a role in the processes of meaning in order to condition and orient the production of meaning. Under these conditions, ethics can be defined as "a mode of behavioral regulation that comes from the individual and that emphasizes co-constructed and shared values to give meaning to his decisions and actions, thus appealing to his personal judgment and responsibility" (Boisvert *et al.* 2003). As Bensamoun and Loiseau (2017) asserted, "Ethics is expected to combine objectives that are as broad as they are vague, ranging from respect for fundamental rights to the dissemination of flexible rules that promote responsible behavior and enable most issues (philosophical, political, legal, economic, educational, etc.) to be addressed in a cross-cutting manner". This innovative method of regulation reveals the virtues of the ethical approach in the field of digital technology: "The advantage of flexibility to achieve objectives in a changing field that would be constrained by rigid rules, which could slow down innovation; attachment to common values to develop framework principles that can be disseminated in an international environment".

In this book, our framework of thought takes its source and inspiration in part from the classical theories of ethics that we are used to browsing. That is to say, the Greek model of virtue[1], where ethics is primarily interested in the individual (the agent) who performs an action, or in the so-called "relational" theories (such as utilitarianism[2], contractualism[3] and deontologism[4]) whose major concern is the nature and moral value of the actions performed by the agent. Our reasoning is based more globally on an ethics oriented toward the people who create or receive the action involving Big Data and undergo its effects.

To do this, we apply universal principles that are both consensual and regulatory, and that tend toward social cohesion. In ethics, the principle

1 Moral principle of trying to be virtuous and using universal casuistic questioning.

2 Universal moral principle of maximizing consequences.

3 A universal moral principle that affirms that every society is based on a social contract.

4 Universal moral principle of categorical imperative (Kant). This ethical theory asserts that each human action must be judged according to its conformity (or non-conformity) to certain duties. It is centered on the respect of rights and duties.

constitutes the foundation that is represented "under the figure of a commandment" (Le Coz 2007). It is immutable, universal, intangible and its value is not influenced by the course of history. This is why all societies tend toward this universality illustrating the uniqueness that surrounds us. The universal is present in the multiplicity of things and, therefore, of the human being. The term principle comes from the Latin *principia*, itself borrowed from the Greek *arche*, which has two meanings:

– first, it refers to "what comes first, what is at the source" (Le Coz 2007). We return to the origins of cultural architecture, to the foundations of the morals, rules of law, customs and habits of a given society;

– second, it means "that which is authoritative" by referring to the "prince" who "comes first" and is vested with supreme legitimate authority.

If we consult the international bioethics literature, we see that four constants constantly recur depending on the country. Thus, references to the principles of "autonomy", "beneficence", "non-maleficence" and "justice" (Beauchamp and Childress 2001) appear over and over in all books, regardless of one's place of origin, culture, beliefs, philosophy or religion. Digital ethics is, therefore, first and foremost an applied ethics, which will have to adapt or reinvent itself depending on the contexts and technologies in place. Thus, AI ethics is a branch of digital ethics specific to robots and other artificially intelligent agents (robot-ethics[5] and machine ethics[6] or ethics of systems). To develop an ethical reflection does not mean, therefore, to act in order to make an ideal reality, but rather to seek to understand the reality so that this understanding modifies our way of being and our way of apprehending this reality in order to adapt to it without submitting to it, so that the evolution of our way of being and acting can transform this reality and modify it.

2.2. General ethical principles

These ethical questions apply to all forms of AI, whether physical robots (such as assistance robots or autonomous cars) or software-based AI (such as expert medical diagnostic assistance systems, intelligent personal assistants,

5 Human ethics to support the elaboration, construction and use of artificially intelligent beings.

6 The ethical rules or behaviors included in the functioning and decisions of artificial moral agents.

or algorithmic chat robots like chatbots). In order to create autonomous machines that improve human well-being and benefit society, it is essential that their design methodology integrates moral values associated with ethical principles. Such a vision oriented toward respect for people will inevitably lead to a change in the current approaches to AI development for organizations. Thus, the ethical reflections associated with these sociotechnological systems must:

– embody the most important moral ideals of human rights;

– to give priority to the optimal benefit for humanity and the natural environment;

– reduce and anticipate the risks and harmful impacts of AI on humanity.

Therefore, in the face of this problem that surrounds the management and use of AI, the quality and harmony of the latter depends on the hierarchy of purposes. This is why we must face certain dilemmas such as providing a quality of meaning and purpose in the accessibility of digital data while ensuring the protection of personal data. AI does not have to define itself as a source of ethics, because it exceeds it. However, it can tend to it and sanction unethical actions. This questioning goes back to the very foundations of ethics. Are there universal values that impose themselves on everyone? Can the singularity of individual knowledge be transposed to a transversal instrument with a collective and multidisciplinary vocation, like an autonomous system? It is on the basis of the following questions that our reflection will be based in order to develop an intellectual approach of questioning with a focus on an ethics of ends and means, and the moral foundations of AI.

When we browse through articles and books on ethics, we quickly become dizzy with the abundance of references and underlying social values used by the authors: "well-being", "quality of life", "pleasure", "happiness", "concern for others", "compassion", "empathy", "solicitude", "altruism", "responsibility towards others", "community solidarity", "sharing", "mutuality", "interdisciplinarity", "precaution", "respect for others", etc. This multiplicity of social values of all kinds has the disadvantage of creating confusion in people's minds. How can we sort through all these values in society in order to disengage from the principles that group them together and allow for a clear discussion and analysis on the subject? What are the guiding principles and rules, the variables of adjustment whose ethical values would be accepted by the community as a whole? Finally, "how can a medical imaging company bring added value and efficiency to patient care?" (Bonhomme and Pinaudeau 2007).

If we take the example of care management, a medical decision must be rational and must be methodically constructed by a few simple, coherent, clear and predefined ethical principles. The notion of assistance has emerged in humanism. This social dimension of care allows patients to regain a part of their autonomy by taking them out of their passivity and by mobilizing their resources. We are, therefore, in a relationship where the person is at the center, in keeping with this one, and where the socioeducational and sociocurative approaches must be developed in the same way as the sociotherapeutic action. It is in this sense that the social environment will restore its hospitalizable value.

As we have seen previously, the term "principle" is intended to give broad guidelines for action and to set attitudes. It designates a fundamental orientation that inspires action. For a very long time, philosophers had the objective of "reducing all moral requirements to a single principle" (Ogien 2007), taking inspiration from their glorious predecessors, Aristotle[7], Kant[8], Bentham[9] and Mill[10]. Today, our modern society has several major principles that are both stable and few in number. Two books are references on the subject of ethical principles in health: *Public Health Communication Interventions* (Guttman 1996) and *Principles of Biomedical Ethics* (Beauchamp and Childress 2001). The latter book was the first to establish and identify these four main ethical principles. According to Beauchamp and Childress, the place of ethics is that of conscience, questioning and dissent, which must be framed by these four universal fundamental principles. This universality seeks to reveal itself to us through our concrete and practical achievements.

It was during the drafting of the Belmont Report that, for the first time in North America, these four major principles of biomedical ethics (respect for the autonomy of the individual, beneficence, non-maleficence, justice) were formalized. However, as soon as they were formalized and even before they

7 The supreme good.

8 Good will.

9 The well-being of all.

10 The harm principle.

were defined, the authors of this report emphasized that these principles represented a framework for reflection whose application was not self-evident: "These principles cannot always be applied in an indisputable manner to resolve particular ethical problems. The objective is to provide an analytical framework to guide the resolution of ethical issues arising from research involving human subjects"[11] (Amann and Gaille 2007).

According to Lazare Benoroyo, "these ethical principles of bioethics – drawing their sources in part from outside the field of Hippocratic ethics – have tended to devitalize the links between ethics and medicine and to call into question the legitimacy of the cardinal ethical aim that traditionally guided the production of health care"[12]. According to Beauchamp and Childress (2001), consideration of the particular case makes it possible to give the principle a precise meaning without which it can be difficult to apply it correctly. According to the authors, "specification (by the case) is a way of reducing the overgenerality of a standard, giving it a greater capacity to guide action, while keeping it in line with the moral meaning of the original principle".

Thus, whether we look at the recent history of North American medical ethics or at the older history of ethical philosophy, articulating the case and principle has always been affirmed as necessary and fruitful for reflection of an ethical nature. These principles, therefore, impose a moral obligation, but they leave room for the creativity of the actors in the search and elaboration of solutions without imposing a canonical scale of values. As they are not *a priori* hierarchized among themselves, it is healthcare professionals who will have to prioritize them in the search for solutions to conflict situations. In this sense, this approach leaves room for the freedom and creativity of the healthcare providers.

Moreover, according to Le Coz (2009), these four cardinal principles simply play an identifying role, allowing the discussion to avoid making a mistake. In no case are they intended to solve all ethical problems. A principle formalizes a value of an "intuitive, subjective, and imprecise" nature by giving it a verbal outline. The principle will allow the value to give a readable and easily shared meaning in a discussion or analysis. Ethics is

11 Belmont Report.

12 Interview by Jean-François Mondot (2011). *Les cahiers de science & vie,* 121 (February–March), 114.

not imperialist; it is rather the means to answer the question of "how to live together". Establishing principles helps to bring order and coherence to social values.

However, it is important to nuance the universal aspect of these four ethical principles in terms of their applicability. Indeed, as Le Coz (2009) pointed out: "Each State is referred to its history, its culture, its morals, and it would be violently unrealistic to try to wrest it from them in the name of an abstract universal. This is why it does not seem conceivable to impose universal norms by obliging this or that state to respect all demands [...] States cannot simply model themselves on one another. They are always returned to the solitude of their choice, forced to take decisions of which none, in the end, fully satisfies us, since there is always a value that is subordinated, if not sacrificed, to another, as we see with the problem of gamete donation. Each State must determine itself according to the values to which it is most attached". It should then be noted that, in particular contexts, tensions and conflicts may arise between these principles from an individual and societal point of view and vice versa. There is, therefore, no way to deal with such trade-offs.

Principle 1: beneficence	
Vocations	Beneficence contributes to individual and collective well-being. It must comply with two very specific rules: the action undertaken must be beneficial and useful, i.e. have a positive cost–benefit ratio. It preserves human dignity and ensures the preconditions for life on our planet by preserving a good environment for the future of our future generations (sustainability of humanity). This principle is directly associated with respect for human dignity, which implies that individuals are treated with respect as persons, rather than simply as concerned parties.
Questions about AI	How can it be ensured that AI does not infringe on human rights? Once algorithms start to overtrain the human being (especially in medicine via medical diagnosis in dermatology and/or radiology), is it unethical not to use algorithms? How can it be ensured that AI well serves and protects the physical and moral integrity of humans, as well as their personal and cultural identity?

Principle 2: autonomy	
Vocations	Autonomy refers to the fact that a person gives himself or herself his or her own rule of conduct, since the Greek terms *autos* and *nomos*, respectively, mean "self" and "law or rule". The purpose of this principle is to involve the patient in the decision-making process. Taking decisions autonomously without outside influence (free and informed consent and free will). With AI, the situation becomes a bit more complex; when we integrate AI in a human practice based on intelligence, we inevitably give up part of the decision to machines. Thus, it seems essential to find a decisional balance between human and machine. For this, we must promote the autonomy (moral, functional, political) of all human beings and the control of the autonomy of the algorithmic system. In order to guarantee human intervention, mechanisms should be put in place to ensure responsibility and accountability.
Questions about AI	How can we make sure that AIs are responsible and maintain the individual freedom of individuals? How can we ensure that human beings interacting with AI systems can maintain complete and effective self-determination over themselves? Often judged infallible and "neutral", do they not open the way to excessive trust and the temptation for everyone not to fully exercise their responsibilities? How can we deal with the new forms of dilution of responsibility that AI implies? How can we be sure that AI protects against direct or indirect surveillance, coercion, manipulation or deception? Should AI be "punished" for actions that are bad for humans and the environment by imposing sanctions? (Legal and moral status of the machine.)
Principle 3: non-maleficence	
Vocations	The objective of non-maleficence is to avoid harming and hurting the person for whom one is responsible and to spare him/her from harm, suffering, damage or even physical, psychological, financial or social consequences that would be meaningless to him/her. AI-specific harms may arise from the processing of data about individuals (i.e. how it is collected, stored, used, etc.). Its purpose, therefore, implies that one does good and refrains from doing harm. This principle appears in the Hippocratic maxim *primum non nocere*[13],

13 "First, do no harm".

	whose consequence is to do good to patients and to keep them away from evil and injustice. Privacy is characterized as being intimately linked to access and control over the use of personal data. Under these conditions, confidentiality, privacy, security, protection and precaution are notions directly associated with this principle of non-maleficence. Inclusion and sociocultural diversity are key ingredients of harm prevention to ensure that these information systems are appropriate for all cultures, genders, ages, life choices, etc.
Questions about AI	How can we make sure that AI is transparent and explainable? How can it be ensured that AI avoids any negative discrimination, manipulation or profiling? How can AI protect society from ideological polarization and algorithmic determinism? How can it be ensured that AI does not have negative impacts and consequences on the environment? (Eco-responsible behavior and sustainable environment.)
	Principle 4: justice
Vocations	Justice is about sharing available resources among all people[14]. This principle is closely linked to the notions of equality and equity that are directly involved in the process of a judicial decision. Its mission is to promote prosperity and preserve solidarity. Ideally, all actions should aim at perfect equality, but depending on the circumstances and the nature of the people, equity is often required in order to establish priorities and a certain hierarchy in the actions to be carried out. This principle has a scope that can be described as "macro-ethics" concerning all citizens, whereas the three previous principles have a much more individual and relational dimension that is considered "micro-ethics". This principle seeks to remove all types of unjust discrimination. The use of AI creates inputs, assets and benefits that are shared (or at least shareable) avoiding the development of further social damage and degradation. Equality also encourages the appropriate respect of inclusion of minorities, who are traditionally excluded, including workers and consumers. Finally, this principle of justice also requires those who develop or implement AI to adhere to high standards of accountability.

14 Resources in time, money or energy.

Questions about AI	How can we educate, train and sensitize citizens about the benefits, challenges and risks of AI for society? How can we apprehend and anticipate new issues around bias, sociocultural diversity, discrimination and exclusion? Although segmentation and profiling have certain advantages for the individual, how can AI significantly affect collective logics essential to the life of our societies (democratic and cultural pluralism, risk sharing)? (Algorithmic fragmentation: personalization and individualization *versus* collective and social logics.)

Table 2.1. *Vocations and questions of ethical principles applied to AI*

Finally, it is important to mention the notion of respect for the "human dignity" of AI, which can be expressed as a principle of respect for the autonomy of the user. The main idea conveyed by this principle is that user autonomy must be the central cog in the functioning of an AI system. First of all, this implies that humans must always be informed, if necessary, that they interact with an AI system. Second, it implies that they must be equipped with the knowledge and tools necessary to interact with an AI system. Finally, it requires that AI systems are created to complement or even reinforce the cognitive, social and cultural skills of humans; in other words, they must be created to serve the development of the human being.

Furthermore, respect for human dignity is also broken down into a principle of human supervision in design. It is a question of setting up mechanisms of governance of AI systems that reconcile the Human-in-the-Loop (HITL) approach, the Human-on-the-Loop (HOTL) approach or the Human-in-Command (HIC) approach. HITL consists of ensuring human intervention in all decision-making by an AI system, but the High-Level Expert Group on Artificial Intelligence considered it not really feasible. HOTL is the assurance of human intervention in AI system design and HIC is a general human supervision of AI system activity. Such mechanisms make it possible for example to determine, before launching an AI system, the use cases in which it is better to refrain from using it.

2.3. Problems and ethical issues specific to the digital environment

Ever faster, ever lighter and more varied, information has an exponential growth that seems to double every 2 years. This is why NICTs are investing in the different fields of society and must allow both qualitative and economic optimization of all industrial and civil processes, while adapting to the regulatory and legislative framework or making it evolve. Thus, the technological progression and popularity of the Internet have led to a gigantic capacity to collect, process, combine and store data. Calculating statistics from these data has enormous potential value for our society, but these benefits come with significant risks, particularly for the privacy of the "owners" of the data. The creation, maintenance, storage, distribution and use of Big Data then become a major challenge. This digital shift is based as much on collaborative networks of citizens and/or scientific researchers, as on cloud computing, or on connected devices and sensors, enabling each user to have access to more targeted and detailed information. This digitization has also been developed on large computer memories and on the algorithmic processing of Big Data, encouraging the production of more powerful tools, particularly for decision support.

On the other hand, it is noted that certain first-order problematic situations associated with AI can lead to ethical tensions:

– the specification of AI that cannot fully and correctly integrate the definition of a concept in natural language, hence the inherent inaccuracy that leads to misinterpretation;

– verification of AI whose learning cannot be proven to follow a predefined framework in all situations, which raises questions of accountability;

– the instability of AI learning that will not "humanly" or "correctly" classify data that was not part of its learning corpus, leading to multiple security problems.

From then on, the "algorithmization" of the world affects all sectors of activity, resulting in information on everything that can exist, anywhere, at any time, in order to be collected, cross-referenced and used for a person's benefit or not. All the ambivalence and uncertainty of the phenomenon is expressed in this observation. Individuals are constantly torn between the hope of a beneficial advance and the worry of a personal nuisance. This feeling of fear and mistrust is amplified as soon as it concerns personal data, especially in health, because its nature is directly associated with the person's intimacy. Indeed, according to several studies and surveys[15], more than 85% of French people say they are concerned, and even worried, about the protection of their personal data and the use that can be made of it, particularly under the economic, ethical and legal pressure of the market. Generally speaking, the behavior of a certain number of public and private players with little respect for the rules on security and the use of NICTs is at the root of a growing loss of confidence among citizens. Indeed, certain security services, whose three main strategies, which guaranteed the protection of privacy – namely, anonymization, free prior and informed consent and opt-out – are outdated. Thus, unsecured data processing by subcontractors may lead to re-identification of information or result in data transfer to a third country outside the European Union (Béranger 2016).

> OPT-OUT.–
>
> Opt-out is a term in the legislative and marketing field that describes an e-mail address that opts out of a mailing list. It is also referred to as "permission marketing". Opt-out is allowed in France, if it concerns e-mailing concerning commercial activities between B2B companies. Opt-out represents the methods making it possible for Internet users to stop a mechanism that they consider intrusive or disturbing for their private life, like blocking cookies in order to no longer receive advertisements.

De-identification-based techniques have been found to be significantly ineffective against powerful re-identification algorithms based on background knowledge and correlation between data. An example is a Massachusetts hospital database that was de-anonymized using a public voter database (Sweeney's Attack). This suggests that beyond sensitive and identifiable data, there are also "quasi-identifying" data.

If we take the examples of Kostas Chatzikokolakis[16] , we can see that 87% of the U.S. population can be identified by date of birth and zip code;

15 Sources: Institut CSA (2014). Les Français & la protection des données personnelles. Study, Orange, February; survey commissioned by Lima in June 2015; Institut Toluna (2015). Study, Havas Media, September.

16 Conference by Kostas Chatzikokolakis entitled *La vie privée dans le monde des données*, at the 2015 edition of Horizon Maths organized with IBM France on the theme of health and data. This event was held on December 14 and 15, 2015 at IBM France.

on Netflix 68% of users can be identified from two rankings in the top 500 titles; the Target chain of stores knows from 25 products ordered whether the person is going to give birth or not.

In addition, the algorithm can tend toward discriminations that can be made, on the one hand, by simple cross-checking of data sets, and, on the other hand, by reintroducing the profiling without the need for prior identification. For example, it has been shown that in the United States, algorithms that calculate the risk of recidivism of prisoners and decide whether they should be released on parole or remain in prison maintain discrimination (O'Neil 2018). The data used are biased from the outset, reflecting the fact that the machine does not go beyond human subjectivity, but industrializes it.

Moreover, qualifying and proving discrimination remains a complex issue to be dealt with and depends on legal frameworks, cultural customs and also on the sectors concerned such as access to employment, credit, insurance, health, and so on.

There are two distinct methods for evaluating the discrimination of an algorithmic system:

– testing[17] is in principle used to qualify intentional individual discrimination: sending two very similar CVs that differ mainly only in the modality (gender, ethnic origin) of the sensitive variable tested. This individual strategy has been applied in particular in economics and sociology (Riach and Rich 2002);

– the disproportionate effect (adverse or disparate impact [DI])[18].

Algorithmic discrimination can be expressed if an algorithm is trained on biased data; it reproduces very faithfully these systemic or societal biases.

17 This is the official approach used periodically by the *Direction de l'animation de la recherche, des études et des statistiques* (DARES) and validated by the *Comité national de l'information statistique* to assess discrimination in hiring by large companies in France.

18 The DI is defined by the ratio of two frequencies: the probability of a favorable decision for the group considered to be disadvantaged by the law on the same probability for the favored group. Schematically, if this DI ratio is smaller than 0.8 (the so-called "4/5ths" rule), discrimination, even if unintentional, is deemed statistically significant and the company is required to justify it on economic grounds in order to avoid a conviction.

More seriously, it may even amplify them. Empirically, we can identify three levels of discriminatory biases that must be taken into account as a priority:

– prediction error rates (for example, if a group is under-represented in the learning base, decisions about it are likely to be less reliable). This is usually the case in facial recognition;

– disproportionate effect[19]: a reflection of the social or population bias by which a group is historically disadvantaged (for example, women's income);

– asymmetric errors[20]: these include more false positives and fewer false negatives.

Numerical data used by AI applications may be biased or drawn from incomplete databases or may reflect human discrimination. Biases may be present at all levels of design and launch (including usage) of algorithmic systems. The way AI devices and algorithms are constructed can also integrate biases or lead to inequalities in treatment, or even systematize or amplify them. A system can be biased if the incoming data are biased, or depending on the way the characteristics of the incoming data are linked and classified during the processing step. Bias can also be caused by the way in which the data output is interpreted.

Moreover, most algorithmic biases are not "top-down biases". They do not arise from the precepts and views of the programmer and how these precepts and views affect the choices he or she has made in constructing the algorithm. It is clear that such top-down biases exist and are a big problem. However, "upward biases", not only in AI, but also in the human mind, pose an equally big problem. These types of biases come from the incoming data, their processing at the lowest level, the patterns that the device detects in the data, and the classifications and correlations that it establishes.

Box 2.1. *Biases in numerical data*

19 Highlighting this bias raises obvious technical and political questions. Would it be politically opportune to automatically introduce a share of positive discrimination in order to reduce social discrimination? (Fair learning)

20 This criterion (odds ratio comparison) has been at the center of the controversy surrounding the COMPAS evaluation of the prevention of recidivism risk in the United States in 2016.

The different types of discrimination reflect the importance taken by AI, including learning algorithms, data and the quality of their representivity. What kind of data for what purpose? This question is *de facto* a prerequisite for the creation of an AI application. Under these conditions, the risks and discrimination biases must be carefully evaluated very early in the database development and learning process in order to be corrected at the risk of no longer being able to compensate for them (Fairness by Design).

To this, we can add the risks of locking up Internet users through personalized services according to their behaviors, habits and expectations. Such a phenomenon goes against the free will and empowerment of the person, and participates in a reductive homogenization of information. One of the risks that surrounds the digitization of the public sphere also concerns the fact that people are judged not on their actual act or fact, but on their propensity to carry out a behavior that their personal data lends them according to their relationships, desires, needs and habits. This would be tantamount to reducing human potentiality to statistics and probabilities controlled by algorithms. The misuse of algorithms also contributes to amplifying the risk of excessive confidence in the choices recommended by the machine's calculations, which are based on potentially erroneous or random assumptions, but which are likely to influence people's choices without them being really aware of it. From a more macroscopic point of view, we are entitled to ask ourselves the question of the "solutionist" risk of a society that systematically turns to an algorithmetic approach that partly masks the complexity of the socioeconomic issues that require other types of intervention. Some algorithmic systems have a significant social impact; for example, they can massively influence political behavior through "filter bubbles". Indeed, algorithms can have several effects that directly or indirectly impact citizens (see Box 2.2).

Transformative effects that format our representations: With the evaluation via websites of services or products (benchmarked by PageRank algorithms).

Normative effects: By looking at the NICTs of the "good practices or habits" brought by the algorithms integrated in the connected devices, smartphones or platforms.

Conformist filtering and confinement[21] effects: By reducing our ability to uncover and understand something else because the algorithm categorizes us in a well-defined sociocultural class (selection and suggestion of music proposed by Deezer's algorithm).

Individualization effects: Made possible by algorithms that allow for personalizing profiles or more targeted and oriented marketing because of Big Data.

Inequitable and hierarchical effects: For example, differentiating rates, credit rates and insurance premiums based on individual profiles (using dynamic pricing strategies).

Choice manipulation effects[22]: When several Internet users perform the same query on the same search engine, we observe that the results can be divergent due to the fact that personal criteria are incorporated into the ranking algorithm.

Unpredictable effects: Related to the poor quality and bias of the input data that can impact the purpose and result of the algorithm.

Box 2.2. *Effects of algorithms on individuals*

DYNAMIC PRICING.–

The term "dynamic pricing" refers to pricing methods by which prices evolve dynamically and possibly in real time according to numerous parameters.

Moreover, one of the important ethical issues of Big Data concerns their apparent objectivity in illustrating social reality with scientific authority and technical rigor. However, there is a certain obscurantism (even a lack of transparency) in the algorithms used. To this, one can associate results that are susceptible to be distorted by multiple methodological biases. Indeed, the algorithm simply and "stupidly" learns what it is asked to learn, and replicates stereotypes if they are not sorted and classified. For example, the data available and used can be biased through their construction or the choice of questions

21 The algorithms of social networks like Facebook create an informational "bubble" of filtering and confinement, limiting access and exchange with people of different opinions.

22 In 2015, a study by Epstein and Robertson showed that a search engine had the capacity to vary by 20% the voting intentions of undecided people for a presidential election based on the ranking of information that is visible during a query (Balagué 2017).

to obtain them (sourcing). The algorithm itself by its operation and purposes may reflect the biases of its designer, commercial interests, political opinions or even introduce discriminations or exclusions prohibited by law.

In addition, for several years now, we have been seeing the massive distribution of fake news via algorithms, particularly from social networks. In 2017, the study report "Fake News Machine" (Gu *et al.* 2017) states that today, one can influence, even manipulate, an election for $400,000 by setting up a device for the massive transmission of false information on social networks (Balagué 2017).

FAKE NEWS.–

Fake news is misleading information delivered with the intent to manipulate or mislead an audience. In recent years, the fake news phenomenon has spread on the Web at the expense of Internet users. This false information can be propagated for different purposes. Some items are intended to mislead the reader or to influence his/her opinion on a particular subject. Others are fabricated from scratch with a catchy title to densify traffic and increase the number of visitors to a site.

At the same time, scientists have to fight against problems of obsolescence of the methodology used to obtain results and may need to start from the data initially collected for a reanalysis. It should be noted that the costs cannot be taken care of by individual teams and must be taken into account by research structures that must, therefore, record data over time and allow for their accessibility over a long period of time.

Moreover, some denounce the negative consequences that this power of information – illustrated by Big Data – can generate if it is reserved for a few predominant operators. Others put forward the risks of discrimination (in access to employment, American universities (O'Neil 2018), insurance, access to services, determination of the price of products and services) or censorship that the individual could be the target of because of the algorithmic processing of their data in order to offer personalized services with dynamic pricing that could deprive them of their autonomy and freedom of choice. To this, we can add the worries and even fantasies about a "dictatorship" of data and predictive algorithms requiring a definition of a "calculated human" that would call into question the principle of self-determination, a principle that is, in our opinion, the very nature of every human being (Zolynski 2015).

AI thus poses specific ethical issues. Indeed, learning algorithms, which learn from numerous examples, lack transparency and traceability tools to explain their results. Hence, the expression "black box" is often used to qualify the opacity of certain intelligent systems and agents.

Due to the nature of their architecture and their operationality, it is complicated to understand the internal reasoning process of a machine's learning algorithms, and thus to explain a result that emerges from them. These algorithms thus induce problems of transparency, intelligibility and control. They are inseparable from the data they process and the platforms that use them to provide a given service. The structure of a classical algorithm is determined during its design. It is composed of the parameters and variables of the algorithm and the sequence of steps to be followed in order to achieve a result. On the other hand, the learning algorithms of a machine use data to determine their operating criteria throughout the learning process. They can evolve over time as new data are taken into account and, eventually, as old data are deleted. It should be noted that it is difficult to modify the initial foundations of a reasoning, the associative and repulsive relations between a few initial concepts, but on which the whole structure of the reasoning depends, all possible discursive developments. With the learning of the intelligent machine, algorithms increase their size and improve automatically, thanks to data sets, without manually adding lines of code. Thus, the partitioning between the variables of the algorithm and the data corpuses it processes no longer exist.

From these observations, we can affirm that an algorithmic system is transparent when it allows its user to learn and study:

– all input data (origin and quality) and the digital data flow;

– the architecture of the algorithmic device and the method of learning, inference, etc.;

– processes, perimeter of the unexplainable part and internal steps of operational execution of the algorithm;

– control parameters of how the algorithm executes its steps;

– assumptions about inputs, purpose and objectives, models and systems used by the algorithm;

– capabilities, limitations and conditions under which the algorithm should work;

– justifications and explanations of the results produced by the algorithmic processing;

– recourse in case of error.

Box 2.3. *Learning algorithms and transparency*

In addition, there is a risk of reproducing injustices or discrimination in the learning of the machine. This is why supervision of machine learning is particularly important. Some developers set up evaluation processes specifically dedicated to the issue of neutrality in machine learning, reserving test phases before operational launch.

On the other hand, the same people who are worried also retain an important hope toward the opportunities that these NICTs can bring. In this context, a climate of hot and cold has set in, alternating between infatuation and suspicion toward digital technology. This has the consequence of making the relationship of trust between the players (Terry 2014) very fragile in an increasingly digital ecosystem. Indeed, this issue of trust is becoming crucial for individuals who are concerned about their fundamental freedoms, as well as for companies, both in terms of brand image and reputation, and in establishing a lasting and healthy relationship with their stakeholders.

When we look at the history of humankind, human beings have always tended to divert the use of innovations and technological progress by constantly redefining their needs and fields of action. We can take the well-known example of Alfred Nobel, who devoted part of his life to the study of explosives and in particular to the less dangerous use and safe marketing of nitroglycerin. The researcher accidentally discovered that when nitroglycerin is mixed with an inert and absorbent solid called Kieselguhr, it appears much safer to transport and handle. The momentum was created! Very quickly, the application of this invention went beyond its simple perimeter of use, which was initially intended for stone quarries, to extend to battlefields. As we can see, invention almost always precedes use: "When humans have a tool, they excel at finding new uses for it. Often, tools exist even before there is a problem to be solved, and each one contains the potential for unpredictable transformations" (Nye 2008).

In a context of disorientation of minds and concern about the digitization of our world, it seems fundamental to us to return to the fundamentals of ethics. What are our human and moral values? How can they be inscribed and applied in this digitalized ecosystem? In any case, a balance must be established between protection and innovation, which requires us to regulate the use of data that is valued by Big Data. Based on this observation, we are led to think that one of the ethical challenges of the digital world would be to eliminate the existing gaps between the initial human intention linked to openness, sociability, and the co-production of values for NICTs, and their possible drifts and reversals of meaning that we observe through the practices that are carried out.

In addition, we can distinguish three categories of effects and functions of AI on the individual:

– objectification: it designates the device by which the characteristics and behaviors of the person are transformed into digital data;

– mobilization: it specifically concerns the interaction between the citizen and the digital professionals in charge of an AI and the ability of the latter to mobilize the person from their AI;

– involvement: it qualifies the steps taken at the initiative of the person (commitment).

These three parameters can be cross-referenced with eight sociological categories:

– consumption: it describes a market dynamic in which the individual can behave as a consumer;

– communication: it qualifies the exchange of information between actors (exchange and sharing/concern for others);

– the community: it qualifies the capacity of the actors to make their actions converge in a more or less strong way (preservation of the social link);

– accountability: it qualifies the sharing of responsibilities in the construction of an event or an action;

– anticipation: it constitutes the precaution and the apprehension of all the repercussions, risks and negative consequences of the NICTs on the citizen;

– social justice: it represents the equality and fairness of a given action toward society as a whole;

– dependence: it constitutes the influence of NICTs on the daily life of the individual;

– universalization: it illustrates the global, transversal and common aspect of the nature, uses and impacts of technologies on the individual.

From this, we combine these different inputs to build Table 2.2, which highlights 24 analysis parameters around the person in an environment associated with numerical data associated with AI.

> COMMUNITY MANAGER.–
>
> The job of a community manager consists of hosting and federating communities on the Internet on behalf of a company, a brand, a celebrity, an institution or a local authority.

This systemic approach, centered on the person, must now be better taken into account by organizations and society because it has the merit of representing a tool for analysis and understanding to apprehend the new behaviors of people given the observed and expected developments of AI.

Finally, depending on whether one approaches the question from a technical or social perspective, ethical risks are protean and of various natures, but always crucial. As far as data are concerned, the technical challenges are traceability (proof, monitoring, auditability and sourcing[23]), security (fraud, theft, loss, re-identification[24]), integrity (reliability and epistemological and semantic quality), internal models of organization and formatting (passage from data to information), accessibility (digital divide, social exclusion, rejection and marginalization) and non-selectivity of data collection (to guarantee objectivity and qualitative prioritization).

---

23 The algorithm simply and "stupidly" learns what it is asked to learn, and replicates stereotypes if they are not sorted and classified. For example, the data available and used may be biased through their construction or the choice of questions to obtain them (sourcing).

24 It has been found that techniques based on de-identification have proven to be significantly ineffective against ultra-powerful re-identification algorithms based on background knowledge and data correlation.

From the social point of view, it is a matter of developing an ethical culture (training, education and awareness of the person on the ethical misdeeds around the data), guaranteeing a certain degree of transparency and confidentiality (e-reputation, brand image, anonymization and harassment relative to the knowledge of sensitive information), forging a form of deontological code (including rights [to be forgotten, to tranquility, to rectification, to access, to limitation of processing, intellectual property resulting from the analysis of Big Data, etc.]) and duties (responsibilities and transfer of ownership), as well as maintaining regulation and control (control and mastery of the data), while guaranteeing individual autonomy (free and informed consent and individual freedom of the citizen without external influence).

Transposed to the algorithms and to the actors involved, these risks are, from a technical point of view, their reliability and robustness (on their operation and decisions), the quality of their predictions and the accuracy of their decisions, their degree of protection (security and control associated with the algorithm), their finality, the absence of bias and value judgments (of discrimination and differentiation individual or collective and intentional or not), their explainability (documentation on the operationality of the algorithm), their transparency (on the operation and decisions of the algorithm) and their automation (practices without human intervention from machines, robots and other AI applications).

From the social point of view, it will be necessary to ensure their automation and dehumanization (human–machine relations and the place of the human being in the explanation of decisions assisted by self-learning machines, loss of employment that can lead to depression and stress, the accentuation of imbalances in the distribution of wealth, the transformation of trades and professions), their reliability (in the sense that they are trustworthy, credible and non-discriminatory), that they protect privacy (surveillance, profiling, individualization and identification of the citizen), and that they can be held accountable (accountability: internal procedures, code or ethical charter, ability to account for evidence through documentation).

Analysis category	Objectivation	Mobilization	Involvement
Consumption	Personal data as marketing knowledge: these data can be sold or purchased	Buyers of commercial services	Informed and active consumers
Communication	Personal data as a platform medium: these data can be transferred	Users of services providing information	Communicating people
Community	Shared personal data: these data can be shared	Persons participating in associations, forums, congresses or seminars	Creators and users of social media
Accountability	Traced personal data: the traceability of these data makes it possible to share or delegate all or part of the responsibility	Persons responsible (compliant) by delegation	People responsible for themselves (self-management/autonomy)
Anticipation	Personal data objects of precaution: the good comprehension of these data allows for an optimal understanding of all risks and the consequences for the person	Individuals who anticipate impacts in the medium term	People attentive to all that surrounds them (environment, context, evolution)
Social justice	Fair use of personal data: the use of these data is carried out in an equal and fair manner	Fair persons in relation to society	People who empathize with others
Dependency	Personal data influenced by the digital ecosystem: these data vary and fluctuate according to their interrelationship with the digital environment.	People accustomed to digital communication tools	People dependent on social networks
Universalization	Homogeneous and universal personal data: the language and translation of these data are understood by everyone	Online platform users and social networks	People with knowledge and abilities of a community manager

Table 2.2. *Impacts of AI data on individuals*

	Domains		
	Ecological	**Social**	**Societal**
	Positive effects		
Direct effects	Ecological decision support	Social decision support	Societal decision support
Induced effects	Resource substitution, dematerialization and process optimization	New jobs, telecommuting, increased autonomy, "project mode"	New modes of citizen expression, creativity, agility
Systemic effects	Save functionality, share and mutualization (a), circular economy (b), "intelligent" assets (c)	Empowerment, "uberization" as an accessible form of technological innovation	Collaboration, collective intelligence, economy of contribution
	Negative effects		
Direct effects	Ecological footprint of the digital self	Related psychosocial risks associated with digital	Security (e.g. fraud, theft, loss, re-identification) (Privacy by Design)
Induced effects	Induced consumption (for example, consumables)	"Digital divides", automation, taylorization, dehumanization, discrimination bias (e.g. social exclusion, rejection, and marginalization), job destruction	Threats to privacy and confidentiality (surveillance, profiling, individualization and identification of the citizen), or even individual liberties (free and informed consent), decrease in attention span
Systemic effects	Rebound effects, emerging risks (e.g. dependence on digital infrastructures)	Alienation in front of the "machine", "uberization" as destruction models of solidarity	Commodification of everything, anthropological transformation (e.g. transhumanism), technological dependence, dilution and decline of responsibility

a) Sharing and mutualization: Here, material resources (places, objects, machines, infrastructures, etc.) are shared between a much larger number of users or uses (collaborative economy).

b) Sources of value in the circular economy: It includes maximizing the useful life of assets; increasing the intensity of asset and resource use; multiplying asset and resource use cycles, in loops or cascades; and regenerating natural capital.

c) Sources of value of "smart assets": It includes knowledge of the location of the asset; knowledge of the condition of the asset; and knowledge of the availability of the asset.

Table 2.3. *Positive and negative effects of AI in the socioecological sector*

ACCOUNTABILITY.–

According to France's *Commission nationale de l'informatique et des libertés* (CNIL), accountability refers to the obligation for companies to implement internal mechanisms and procedures to demonstrate compliance with data protection rules.

The beneficial and detrimental effects of AI can be identified and classified across ecological, social and societal domains (see Table 2.3).

Finally, it will also be necessary to manage their lifecycle (complete and traceable data traceability, processing, interpretation and visualization), and to envision an algorithmic governance (regulations, labels and certifications, ecological footprint and sustainable environment) that guarantees free will (risks of human dependence on machine decisions).

2.4. Ethical criteria and better risk assessment of AI-related digital projects

The emergence of AI is likely to mask the development of a societal project, modeled according to their economic interests and the increase of their vision. It is, therefore, necessary to be able to identify this ideology, which presents itself as a "state of the science". This is all the more imperative, since it is structured by an extremely reductive approach to the human being and to life, which has not only ethical, but also social, cultural, political, anthropological and legal repercussions, in the sense of the instrumentalization of the human being.

The main problem seems to be the frequent perception that algorithms are capable of making neutral, non-discriminatory and independent predictions about future events. Thus, it is not the algorithms themselves, but the related decision-making processes that need to be analyzed to determine their possible consequences for human rights. The current scientific discourse revolves around concepts such as human autonomy and individual capacity for action, both of which relate to the right to privacy and informational autonomy, without being consistent with the very notion of privacy itself. A distinction must, therefore, be made between autonomy, on the one hand, and capacity for decision and action, on the other hand. These concepts refer to the human capacity to set one's own objectives and to exercise one's free will, respectively. As such, they may prove incompatible with the use of algorithms and techniques of automated processing. In other words, it may be necessary to broaden the definition of human rights or reinterpret them to protect individual autonomy and capacity for decision and action.

This touches on fundamental and essential ethical issues concerning the use of automated data processing techniques and algorithms that are beyond the scope of this book. How can AI incorporate normative values? Some of the ethical issues surrounding autonomous systems provide insight into the complexity of the question: how is the algorithm supposed to infer the probability that an event could harm a person from a hypothetical situation? Is the number of lives potentially at risk a computational parameter? When is a "good" or "bad" decision made in such a situation, and what are the legal consequences? If so, who is held responsible for a "bad" decision? How can we design an algorithm that would determine the action to be taken, in particular by the self-propelled car, and what would be its limits? What decision parameters should be taken into account? By a variable based on the consequences [25]? The deontology of actions[26] ? Moral values[27] ?

What instant choices should a software-controlled vehicle make as it approaches a collision? Is there more (or less) risk of gender, ethnic or racial bias in an automated system? Are social inequalities only reproduced or are they amplified by automated data processing techniques? This study identifies a number of human rights concerns raised by the increasing role of algorithms in decision-making processes. When algorithms violate human rights, who is responsible? The person who programmed the algorithm, the operator who uses it, or the human being who implemented a decision based on it? Is there a difference between such a decision and a decision made by a human being? What are the implications for the exercise of human rights and human rights guarantees as provided for by established standards, including principles of the rule of law and judicial processes? How can we ensure that

25 In the case of an autonomous car involved in an accident, the variable may be unharmed pedestrian, injured passengers, degraded autonomous car, injured pedestrian, unharmed passengers and unharmed autonomous car.

26 An action is considered acceptable if it is "good" or "neutral". Using the example of the autonomous car, the action of "running" a red light may be considered "bad" in the absolute sense (because it violates the Highway Traffic Act), but "good" if it is to avoid immediate danger.

27 For example, the moral values associated with a self-contained car may be compliance with traffic laws, no harm to property, no harm to persons and protection of persons, in the case of an autonomous car.

our basic standards, values and fundamental rights are always respected by these new technologies?

It should be noted that an ethical code – no matter how coherent it may be – can never fully replace ethical reasoning, since the latter must always be receptive to ever-changing contextual details. This is why the notion of Ethics by Evolution seems to us to be a determining factor in the hope of moving toward an AI that is both ethical and eco-responsible.

This book should, therefore, not be seen as an outcome, but rather as the beginning of a new open discussion mechanism.

Before unilaterally setting ethical limits on AI components, we must take a step back and ask ourselves: Is ethics calculable in Turing's sense?

> TURING.–
>
> In theoretical computer science, a Turing machine is an abstract model of the operation of mechanical computing devices, such as a computer. This model was imagined by Alan Turing in 1936 in order to give a precise definition to the concept of algorithm or "mechanical procedure".

The purpose of this chapter is to list and define the main ethical criteria that will constitute the basis of a future reference system in order to move toward AI at the service of human intelligence and thus prevent and anticipate the drifts and possible consequences of the improvement of algorithmic systems.

Thus, if an ethical problem arises, it may follow, for example, the following path:

– take a step back: it is a matter of distancing oneself from one's daily life in order to better evaluate and put into perspective various elements (social, environmental, divergent interests, etc.) before acting solely on the basis of economic and commercial logic. It is also a question of having the capacity to anticipate enviable events, and more modestly to focus on the medium and long-term consequences of our present actions. For example, a developer of AI algorithms can rely on tools to limit the bias of its solutions as soon as they are developed, because it has analyzed, upstream, the risks associated with the use of its device and processes;

– choosing the "best" possible ethics: there are always several possible choices in ethical reflection. This is why we often speak of an "ethical dilemma". Ethical choices often confront different ethical theories (consequentialism, deontology, ethics of human virtues and values, etc.), which can contradict each other, depending on situations and contexts. In ethics, there is never an absolute and definitive answer, but more certainly, there are "best" possible choices, depending on the case. For example, developers may encounter a conflict between their deontology and their personal ethics (or "ethics of virtues") in their professional career.

Part of our research work has allowed us to develop reflections on the design of an intelligent platform for ethical regulation. To do so, we are based among others on systemic approaches such as the work of Ying and Colloc (2015, 2018), Merise method[28] related to the design of an information system (IS), as well as current scientific research in neuroscience and robotics.

MERISE METHOD.–

Merise is a modeling methodology for general use in the field of information systems development, software engineering and project management. First introduced in the early 1980s, it was widely used in France. It has been developed and perfected to such an extent that it has been adopted by most of the major French governmental, commercial and industrial organizations. Merise performs separate data and process processing, where the view of the data is modeled in three steps: from design to physics to logic. Similarly, the process-oriented view goes through the three stages of design, organization and operation. These stages of the modeling process parallel the lifecycle stages: strategic planning, preliminary study, detailed study, development, implementation and maintenance. It is a method of analysis based on the entity-relationship model. Using Merise, one can design tables with relationships to create a relational database.

Consequently, the ethical framework around algorithmic processing can be divided into five interdependent and complementary categories of ethics that shape our approach centered on Ethics by Evolution (see Figure 2.1):

28 According to Arnold Rochfeld, this approach integrates the object definition model, the usage model, the requirements model and the architecture model.

– data ethics (DE): it includes identification of the data processed by the algorithm and the exchange of these data (supervision and traceability of data[29]). This is the domain that concerns the nature and characteristics relating to digital data;

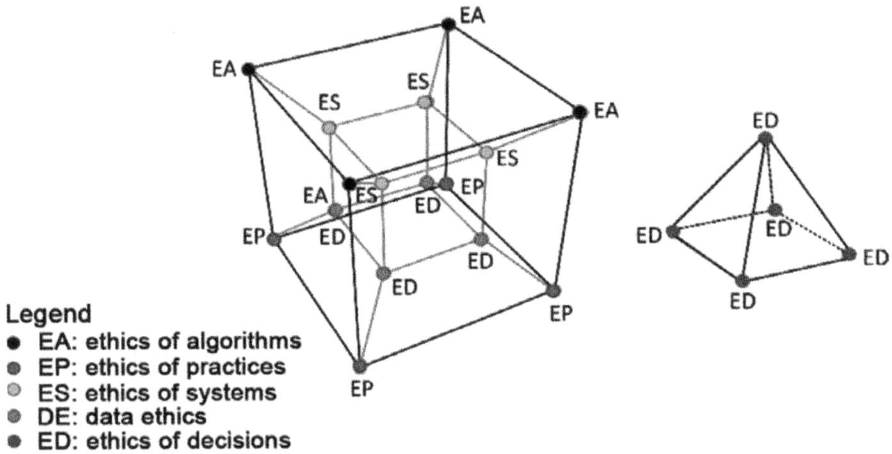

Figure 2.1. *Ethical categories associated with algorithmic processing. For a color version of this figure, see www.iste.co.uk/beranger/responsibility.zip*

– ethics of algorithms (EA): it includes application to operations and processes associated with the exploitation of digital data throughout their lifecycle. In other words, to the functioning of the algorithm on a scale sufficient to understand and explain what it actually does (Black Box *versus* transparency);

– ethics of systems (or machine ethics) (ES)[30]: it includes ergonomics and functionality of the technological architecture of the information system, in adequacy with the skills and practices of the user (artificial moral agents and human-machine interfaces);

---

29 Completeness, accuracy, consistency, timelessness, duplication, validity, availability and source of data must be considered in a thorough analysis. It should be noted that the data may include any kind of bias, whether due to the data used for training or the data captured during the operation of the algorithm. This can occur if the data are incomplete, outdated, duplicated or insufficient. Another category of bias may arise from the design of the algorithm, if the bias is due to an error in the code or false "patterns". Finally, bias may appear in the output data, in the sense that a misinterpreted or misused result, or the assumption of false or incorrect assumptions, may generate risks.

– ethics of practices (EP): it includes explanation of the quality of the results of the algorithmic processing, the purpose and uses of digital data processing (purpose of the algorithmic processing). The practice is based on the use of techniques, the explanation of their purpose, the dismantling of reasoning, and, above all, the embodiment of the interhuman relationship. There is not one practice, but practices. Recommendations are only a theoretical framework from which the mass of all particular cases escape. It is in this departure from the framework, in this "transgression," that the "ethics of practices" is also situated;

– ethics of decisions (ED): it includes societal impacts on both people and the environment (digital humanity and digital ecology). It can be noted that decisions can be revealed in two different temporalities. Firstly, upstream of the design of a digital solution, these decisions are operational in nature. Secondly, after the development of the digital application, these decisions are of a strategic nature.

<table>
<tr><th>Ethical analysis</th><th>Function</th><th>Digital value</th><th>Algorithmic ethics</th><th>Content</th><th>Area of application</th></tr>
<tr><td rowspan="4">Descriptive ethics</td><td rowspan="4">Application and performance</td><td rowspan="4">Intrinsic (design) value</td><td rowspan="4">Data ethics
Ethics of systems</td><td rowspan="4">Means, devices, channels and procedures</td><td>Strategy and methodology</td></tr>
<tr><td>Organization and regulations</td></tr>
<tr><td>Structure and technology</td></tr>
<tr><td>Relationship and culture</td></tr>
<tr><td rowspan="4">Normative ethics</td><td rowspan="4">Regulation</td><td rowspan="4">Management (implementation) value</td><td rowspan="4">Ethics of algorithms</td><td rowspan="4">Standards, codes and rules</td><td>Steering/governance</td></tr>
<tr><td>Shaping</td></tr>
<tr><td>Development cycle</td></tr>
<tr><td>Operations</td></tr>
<tr><td rowspan="4">Reflexive ethics</td><td rowspan="4">Legitimation</td><td rowspan="4">Operating value (use value)</td><td rowspan="4">Ethics of practice
Ethics of decisions</td><td rowspan="4">Principles and ethics</td><td>Beneficence</td></tr>
<tr><td>Justice</td></tr>
<tr><td>Non-maleficence</td></tr>
<tr><td>Autonomy</td></tr>
</table>

Table 2.4. *Structuring of ethics of algorithms*

Ethics, on the other hand, resides in the intention turned towards the finality and meaning of an action. It can be divided into three types of ethics (see Table 2.4):

– descriptive ethics: it applies to intrinsic (design) value. It constitutes an ethics of application and assignment in the form of practice with means, devices, ways, and procedures implemented;

– normative ethics: it concerns the management (implementation) value. It forms an ethics of regulation of deontological aspect *via* standards, codes, and rules put in place;

– reflexive ethics: it applies to the exploitation (use) value. It represents an ethics of legitimization based on the questioning of foundations and purposes through human principles and values.

The articulation and arrangement of these three families of ethics apply to the entire lifecycle of an algorithmic system (design–implementation–use) to feed our Ethics by Evolution.

This approach also allows the evaluation of an algorithmic system according to application sectors, stages of the digital data lifecycle or associated themes and concepts. It is essential to be able to have evaluative mapping from several different viewpoints so that the study can be understood by all the actors directly and/or indirectly concerned by the algorithmic processing. Indeed, with the nature and impact of digital technology being transversal, it becomes a necessity to acquire a multi-sectoral and multi-disciplinary analysis so that each protagonist can visualize and, above all, understand the results from a pragmatic point of view. Ethics must, in this sense, be integrated into all stages of the algorithmic chain so that our fundamental values are respected from the outset. If organizations and institutions have already established avenues for reflection in this field, it is necessary that ethics go beyond this stage of reflection to begin to be applied in practice: it is time to take action by establishing rules that can be followed and respected right away by designers and developers. This is why a collaborative work between the operator of the algorithm, the individual and the regulator must also be set up, and this for several aspects – first, because it is he or she who designs the algorithm, sets the foundation for settings, chooses the learning data and determines the purpose of the processing; second, because he or she must remain in control of the information and must be able to modify, rectify or erase data that are inaccurate or out of date; third, because he or she must exercise the role of support and controller of these technologies, at the same time as guaranteeing the protection of the individual's personal data. Each step of the algorithm must also be documented, in order to better trace any bias that may appear throughout its operation. With regard to possible malfunctions of the algorithm, the implementation of a detection and warning system is also desirable. All this control and monitoring of the algorithm must be validated and revised regularly in order to preserve its accuracy and uniformity over time.

Under these conditions, beyond an analysis by family of algorithmic ethics, we can cut out three different aspects:

– by sectors of application: strategy and methodology, organization and regulation, structure and technology, relationship and culture, steering, shaping, development cycle, operations, beneficence, justice, autonomy and non-maleficence;

– by stages of the data development cycle: generation and acquisition, storage and conservation, analysis and processing, restitution, recovery and integrity of the lifecycle;

– by themes and concepts associated with digital technology: interest and purpose, decision-making, quality of information, contributions (time savings, efficiency gains, etc.), acceptability, accessibility, traceability, security, regulations, relevance and consistency, privacy and confidentiality, transparency and explainability, structuring of information, validation and verification, durability, evolution, interactivity, policy, HR organization, steering and governance, sociocultural diversity, environment.

In addition to direct regulatory mechanisms designed to influence the code of algorithms, indirect mechanisms developed for the same purpose could also be considered. These concern the production process or the producers of algorithms and seek to ensure that the latter are aware of the legal issues, ethical dilemmas and human rights concerns raised by automated decision-making and data processing. This could be achieved through standardized professional codes or forms of licensing systems for data engineers and algorithm designers, similar to those that exist for professions such as physicians, lawyers or architects. The idea of improving existing mechanisms for software management and development processes is also frequently raised. It could be particularly applicable to agile software development techniques, where modularity, temporality and capture pose considerable problems in terms of privacy and other human rights. Since the use of algorithms in decision-making has the potential to infringe on individual rights, additional control mechanisms could help ensure that the algorithm is applied in a fair and sustainable manner.

The objective of such an approach is to contribute to the development of know-how for the responsible and ethically acceptable production, study and use of data, and to disseminate good practices. To achieve this, it is essential to implement a proactive control tool throughout the lifecycle of digital data through this new technological approach of responsible design: Ethics by Evolution. This form of ethics seems to us to be a determining factor in order to be able, in the medium term, to evaluate humanoid and android robots based on their "moral personality" or their "ethical integrity".

Indeed, we can take the example of the protection of personal data, for which we encounter difficulties in dealing, *a posteriori*, with some of its associated problems due to the complexity of the processes. The idea is, therefore, to integrate ethical principles and rules as early as the design of technological tools, in their very architecture. This allows for ethical and social questions to be taken into account as early as possible in the creation and traceability of a technological innovation. This approach therefore imposes the introduction of ethical criteria into the very development of NICTs, i.e. in their applications and/or in their computer source code (see Figure 2.2).

SOURCE CODE.–

Source code is text that represents program instructions as written by a programmer. The source code often materializes as a set of text files. It is usually written in a programming language that allows for better human understanding. The analogy between source code and a recipe is often used. Indeed, a recipe is an organized list of ingredients whose quantities and functions are defined. The goal is to obtain the result desired by the cook, according to a technique and a determined sequence of operations.

This evaluation is made up of 36 ethical criteria divided into five families of ethics that make up our digital ethics.

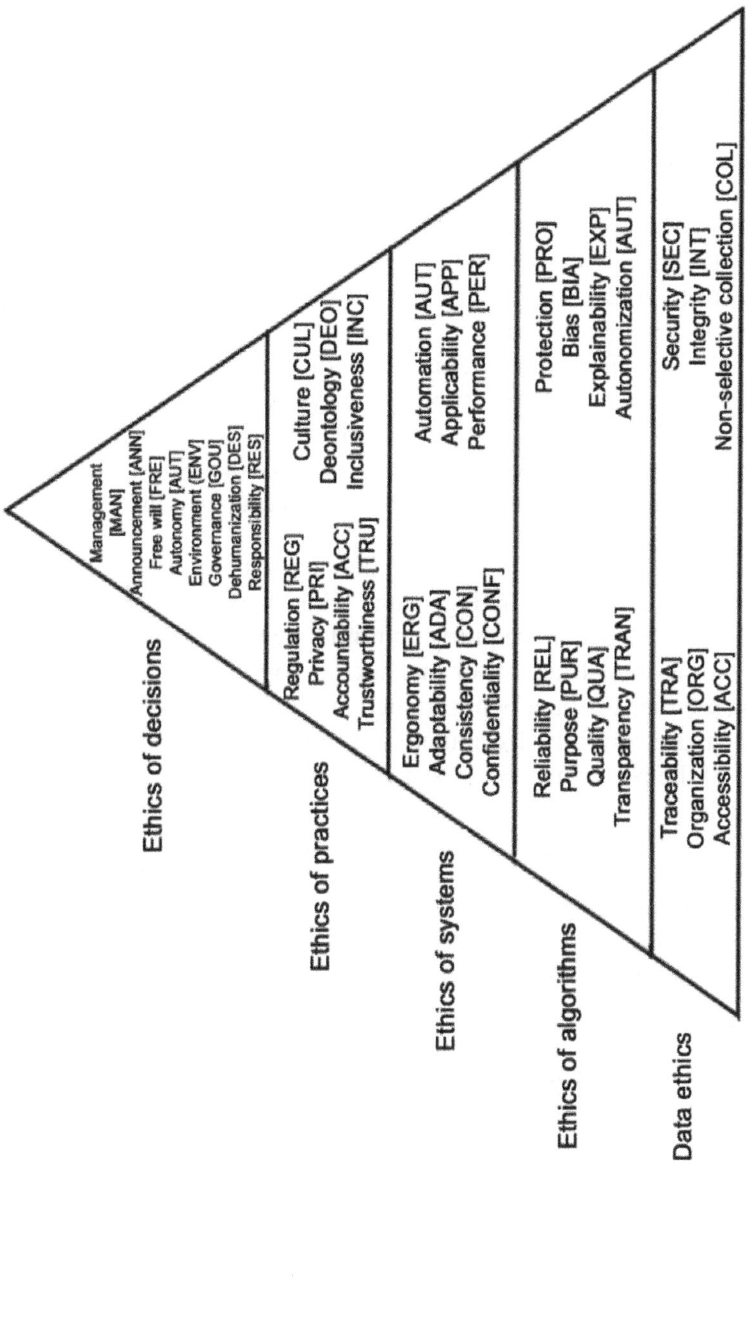

Figure 2.2. *Ethical criteria for algorithmic processing*

2.4.1. *Data ethics*

– Traceability (TRA): It is the situation in which the necessary and sufficient information is available to know (at the time or in retrospect) the nature and composition of the data throughout its lifecycle (from its production to its destruction or recycling).

– Security (SEC): It is the state of a situation that presents the minimum risk. It can be considered individually or collectively, either as an objective, a right, a value (freedom), a state of what is secure and a function or act aimed at securing the processing of digital data. In general, security is closely related to confidentiality, integrity, accessibility and availability around data. Security consists of ensuring that the algorithmic device acts as it is supposed to, without harming users (human physical integrity), resources or the environment. This includes reducing unintended consequences and errors in the operation of the system. Processes to clarify and assess the potential risks associated with the use of AI-related products and services must be in place. Safety devices must be based on Safety by Design mechanisms and take the necessary precautions against misuse.

– Organization (ORG): It defines the action of structuring, delimiting, arranging, distributing or articulating data. This is both a social and transversal process because it requires multiple human skills. It can be carried out according to a structural or managerial approach. It is, therefore, necessary to ensure that the data being trained, as well as the validation and testing of the data, are correctly distributed in the system, in order to obtain a realistic picture of the performance of the algorithmic system.

– Integrity (INT): The state of data that has all its original components without altering its intrinsic qualities. The integrity of the data is analyzed mainly from the point of view of its structure and internal constitution. Indeed, the introduction of malicious data into the information system can modify the behavior of AI solutions. This is particularly true and essential for self-learning systems.

– Accessibility (ACC): The access of data to the individual based on certain criteria and profiles. The issue around accessibility is central to the risks involved in the protection, confidentiality, privacy and security of data.

– Non-selective collection (COL): The basic concept of a Big Data project is to acquire a large volume of heterogeneous data from different sources and of different natures, without prior selection, and to cross-reference them together to see if any correlations emerge. More and more questions are being asked about data cleansing and hierarchical data selection. Indeed, some data sets that serve as a learning base for algorithms can convey cognitive biases. Some databases may, for example, in a particular field, contain a cultural and historical bias in terms of male–female representativeness. It seems important to be able to explain the content of the data selected and used by learning algorithms in order to ensure their neutrality.

It is then important that the data entered into an AI system, on the basis of which it was trained, accurately represent the diversity of society. It is then essential to prevent discriminatory bias from the earliest stages of design, paying particular attention to vulnerable people (women; children; the elderly; the economically disadvantaged; members of the LGBTQIA+ community; people with disabilities; racial, ethnic or religious groups). Concerning AI systems for medical decision support, it is important to ensure that rare cases are well represented in order to avoid diagnostic errors. Although discriminatory biases can be directly sourced in the programming of an AI system, they are most often due to the poor quality of the batch of data entered in the system. These biases are sometimes voluntary (e.g. linked to economic or financial interests), but this is not the most frequent case. Several phenomena can create discriminatory biases: a data set reflecting pre-existing inequalities in society, a data set that is too unrepresentative, or a data set that is fairly representative but which gives rise to false statistical correlations. Such biases, which are already problematic, are sometimes accentuated by the programming of the AI system. In this sense, it is necessary to address the problem of "outliers", which are particularly important in the health field; in algorithms, "outliers" are those that are excluded because they are too far away from the mean to be used in the construction of decision-support options. However, in health, these extreme cases, which AI would no longer be able to take into account in order to reinforce its overall efficiency, always correspond, in practice, to human situations.

2.4.2. *Ethics of algorithms*

– Reliability (REL): It involves the analysis of system failures, particularly from a statistical point of view. An algorithm is said to be "reliable" when the probability of carrying out its mission and function under given conditions over a given period of time corresponds to that specified in the initial specifications. A distinction must be made between reliability (function of time) and quality control (static function). Reliability is based on the robustness and the maintenance process used by information systems. These criteria could potentially be considered as parameters to reinforce transparency and accountability. Reliability requires that the accuracy of results can be confirmed and reproduced[31] by an independent assessment. However, the complexity, non-determinism and opacity of many AI systems, as well as their sensitivity to training and modeling conditions, can make it difficult to reproduce results. One dimension of cybersecurity is to ensure the reliability, in a broad sense, of predictions or decisions calculated by an AI system. This refers to the "robustness" principle of AI systems. First of all, it is necessary to guarantee an optimal accuracy of predictions. This implies, for example, when incorrect predictions cannot be avoided, clearly indicating the margin of error to be taken into account. Second, the creators of AI systems must ensure the reliability, in the strict sense, as well as the reproducibility of predictions. Strictly speaking, reliability implies that with the same data entered into the AI system, the system calculates correct predictions in different situations. Reproducibility, on the other hand, assumes that with the same data entered into AI systems, the system calculates identical predictions in identical situations. AI systems must operate reliably and safely:

- AI systems must be reliable under all circumstances and subjected to rigorous testing to demonstrate that they can operate reliably, safely and continuously (before and after launch);

- this requires the creation of clear frameworks and integrated controls;

- protection against takeovers or the introduction of malicious data must be given increased attention;

- human beings must always be able to decide to stop the system.

---

31 Reproducibility is essential to ensure consistency of results in different situations, computational frameworks and input data.

– Protection (PRO): Securing an algorithm requires a study of the risks to which it is exposed and the choice of technical or organizational solutions that guarantee its confidentiality, auditability, integrity and availability. This serves to protect against malicious acts (theft, sabotage, attacks against an algorithmic device, misappropriation, or deterioration of intangible assets), accidents (partial or total destruction or malfunction of the algorithmic code integrated into a platform, a connected device, an application or a self-learning expert system) and errors (which may occur during the operation of the algorithm). As for cybersecurity, it must be guaranteed throughout the lifecycle of an AI system, so that it also concerns both data and AI systems as such. First of all, the designers of AI systems must ensure the quality of the data entered in an AI system, this quality being essentially based on the absence of bias (*supra*). Second, they must anticipate as much as possible digital attacks that could affect these data, for example, by providing double backups. Third, they must anticipate digital attacks or malfunctions that could challenge the initial purpose of the AI system. To do this, they should integrate a reversibility mechanism or even an emergency stop button into the AI system, the ideal being a continuous monitoring to detect any deviation. In extreme cases, they should ensure that the AI system programming cannot be modified after its creation.

– Purpose (PUR): Its character tends toward a goal, an end in itself. One of the first issues in the study of an algorithm often revolves around the objectives of the algorithmic processing. We look at the initial objectives and the results that must emerge from the exploitation of the data. Those who control algorithms may intentionally attempt to obtain unfair, discriminatory or biased results in order to exclude certain groups of people. Intentional harm may, for example, be achieved by explicitly manipulating the data to exclude certain groups.

– Bias (BIA): By definition, search algorithms and search engines do not treat all information in the same way. If the processes of selecting and indexing information can be used systematically, search results will generally be ranked according to their supposed relevance. Thus, different pieces of information will have different degrees of visibility depending on the factors taken into account by the classification algorithm. Due to data aggregation and profiling, algorithms and search engines rank the advertisements of small companies registered in less privileged regions below those of large companies, which can put them at a commercial disadvantage. Search engines and algorithms also do not treat all users equally. Different users may obtain different results, based on behavioral or other profiles, including individual risk profiles that may be established for insurance or credit purposes or, more generally, for the purpose of differentiated pricing, for example, by proposing different prices for the same goods or services to different consumers based on their profile. Furthermore, errors in the method or non-compliance with protocol rules can generate erroneous results for the algorithm. The data that feed the algorithm may contain biases, particularly discrimination[32], and are, therefore, found in the algorithm itself. A biased algorithm that systematically discriminates against a group in society, for example, on the basis of age, sexual orientation, race, gender, or socioeconomic status, may raise serious concerns not only for the access to rights of the consumers or end-users affected by such decisions but also for society as a whole. The sensitive variable of an algorithmic system must be known. This is obviously not always the case, and this is a problem because removing the sensitive variable from a model does not necessarily avoid a discriminatory decision, but prevents the bias from simply being identified. The other difficulty is that it is also necessary to prove discriminatory intent. However, in the case of discrimination by algorithmic processing, the discrimination is not necessarily the result of intentionality. Other biases can be observed, notably associated with the fact that predictions are systematically elaborated from patterns and sequences observed on a pre-existing database and, therefore, on data from the past. The algorithm will, therefore, tend to perpetuate the same functions, for example, by reproducing a discriminatory character on a cohort of people. These biases are likely to have repercussions of an individual or collective nature, whether intentional or not. Discrimination may be unintentional due, for example, to data problems such as patterns of bias, incompleteness and poor governance. Machine learning algorithms identify patterns or regularities in the data and, therefore, will also track patterns resulting from biased and/or incomplete data sets. Therefore, an incomplete data set may not reflect the target group it is intended to represent. Although it is possible to remove clearly identifiable and undesirable biases in data collection, there is always some bias in the data.

32 Discrimination refers to the variability of AI outcomes between individuals or groups of people based on the exploitation of differences in their characteristics that may be considered intentional or unintentional (such as ethnicity, gender, age or sexual orientation), which may have a negative impact on those individuals or other groups of people. For example, one source of bias in the nature of the data may be associated with age, gender and sexual preferences, which necessarily leads to some sort of categorical exclusion (Diakopoulos 2014).

Hence, early identification of possible biases, which can be corrected later, is important for structuring the AI application.

– Quality (QUA): The internal "way of being", good or bad, of an algorithm. It is defined as the ability of a set of intrinsic characteristics of an algorithm to satisfy requirements that can be translated into the forecast quality or the error rate associated with the use of a learning algorithm. In statistical learning, the accuracy of the decision depends on the quality of a forecast and, therefore, on the quality of a system or an algorithm. The latter depends on the representativeness or bias of the initial data, the suitability of the model to the problem posed and the quantity (variance) of residual noise. It is evaluated on an independent test sample or by cross-validation (Monte-Carlo), but remains indicative in the form of a probabilistic risk of error. Once the algorithm has been trained and optimized on the training sample, it is the estimation of the prediction error on an independent test sample of sufficient size that gives an indication of the quality of an algorithmic decision. The models and algorithms created should also be able to be stored and executed in secure environments, guaranteeing the integrity and intangibility of the device and process.

– Explainability (EXP): The action that implies that the activity and operationality of the algorithm is explainable. Explainability – as a form of transparency – requires the ability to describe, inspect and reproduce the mechanisms used by AI systems to make decisions and learn to adapt to their environment, as well as the source and dynamics of the data used and created by the system. More specifically, we are able to explain and understand the algorithm under study. Causal explanations that link digital experiences to the data on which they are based can allow individuals to better understand how the algorithms around them influence their lives and behaviors. Thus, being explicit and open about choices and decisions about data sources, development processes and stakeholders should be required of all AI systems that use or affect human data or may have other morally significant impact. Transparency in the "reasoning" of an AI system implies being able to trace the "choices" made by the system to arrive at a prediction or decision; this is referred to as "explainability". In addition to the simple logic of transparency, this principle is essential to provide an explanation in case of malfunction of an AI system. Concerning classical algorithmic systems, the principle of explainability can thus be applied by the designer, who will have to detail the programming method of the learning model. However, concerning the diagnostic suggestions and decisions calculated by algorithmic machine learning systems, they cannot be justified other than by statistics. This explainability refers to the need to understand and re-sustain the AI decision-making processes. This notion is closely associated with accountability and transparency. Moreover, an algorithmic decision is interpretable if it can be explicitly accounted for on the basis of known data and characteristics of the situation – in other words, if it is possible to relate the values taken by certain variables (the characteristics) and their consequences on the prediction, for example, of a score, and thus on the decision. On the other hand, an algorithmic decision is said to be "explainable" if it is only possible to identify the characteristics or variables that participate most in the decision, or even to quantify their importance.

– Transparency (TRA): The obligation to report and inform the public about the content and functioning of the algorithm. It is associated with a reduction in the asymmetry of information. Transparency is generally based on three pillars: monitoring (automated or not), reporting (periodic reporting of performance) and verification. In the case of an opaque algorithm, it is impossible to simply relate values or characteristics to the result of the decision, especially in the case of a nonlinear model or with many interactions. Such a high value of a variable can lead to a decision in one direction or another depending on the value taken by another unidentifiable variable, or even a complex combination of other variables. Machine learning techniques further complicate the understanding of algorithms to such an extent that the disclosure of all their source codes may not be sufficient to make them more transparent. It would be preferable for this to be done by explaining precisely how the results of an algorithm are produced. Since algorithms may cast doubt on the fact that they have led to decision-making, and since their examination is only of interest if it is conducted within their social and organizational context, it might be interesting to focus measures to promote transparency on the decision-making process itself. AI systems need to be understandable:

- transparency and accountability are necessary for the proper functioning of AI;

- people need to be able to understand the basis for decisions that affect their lives;

- it is then not enough to simply publish the algorithms, but also to explain them.

– Autonomization (AUT): The process that allows a structure or system to acquire the means necessary for its autonomy. NICTs such as self-learning expert systems, AI or autonomous cars are the illustration of this self-naming of machines.

2.4.3. *Ethics of systems*

– Ergonomics (ERG): The infrastructure and architecture of the algorithmic processing is adapted to its user and work environment. Moreover, while capturing the customer's attention and seeking to keep it for as long as possible is a well-known marketing strategy, and often quite visible from the outside, it should be noted that this type of strategy is less perceptible with digital tools because the techniques for capturing attention are subtler, and can moreover make use of playful mechanisms. Some interface design tricks designed to deceive the user are called dark patterns. AI applications should not have a one-size-fits-all approach, but should be user oriented, taking into account the user's skills, abilities, needs and requirements. Conceiving a design and ergonomics for everyone, at all times, is one of the guarantees of their good inclusion, whatever the contexts and situations. The design of an AI system refers to the interface between the system and its user. The designer of an AI system can, for example, surround himself or herself with a team that is representative of the target audience, or ensure that the system's parameters can be adapted to a wide range of individual preferences.

– Adaptability (ADA): The machine has a functional flexibility that responds to both the requirements and expectations of its user. AI applications must be built in such a way that every person can use them, regardless of their nationality, age, disability status or socioeconomic status. This adaptability can be achieved over time, since digital processing is evolving according to changes in its environment and the needs of its user. Adaptability is highlighted in self-learning algorithms that use data to create original patterns and knowledge and generate decision rules through automatic learning techniques. By adopting various learning styles, algorithms can model problems from data sets and propose new solutions that would be impossible for a human being to grasp.

– Consistency (CON): This translates into the usefulness, interest and meaning that the algorithmic system has through its purpose and operation.

This coherence must meet, on the one hand, the expectations of citizens, and, on the other hand, the requirements of professionals involved in AI.

– Automation (AUT): The total or partial execution of a task, a function and a sequence of technical operations in an automatic manner by machines operating without human intervention. This phenomenon is one of the cornerstones of every industrial revolution that humanity has known. Automation is one of the main characteristics associated with algorithmic decision-making. The ability of automated computer systems to replace human beings in an increasing number of situations is an essential feature of the practical implementation of algorithms. The replacement of humans by automated computer systems generally has its origin and justification in old problems such as large-scale data processing, speed and volume of decisions to be made and, in many cases, is based on requirements for lower error rates than those observed in humans.

– Confidentiality (CONF): It involves ensuring that data are only accessible to those whose access is authorized. It is one of the cornerstones of information security. AI systems must be secure and respect privacy:

- data confidentiality is a pillar of AI system development;

- compliance with data protection principles and laws must be built into AI systems while allowing the system to function and provide satisfactory service;

- to this end, techniques for separating personal data from information data may be implemented.

– Applicability (APP): NICTs must be easily usable and applicable by users. For this, the designer must pay attention to the functional and pragmatic side of the technology.

– Performance (PER): The performance of an information system can be part of a ternary relationship between the objectives sought (targets, estimates, projections), the means to achieve them (human, material, financial or informational resources) and the results obtained (goods, products, services, etc.). Performance measurement is then carried out on three axes: relevance (the relationship between the initial objectives and the resources acquired to achieve them), efficiency (the relationship between the results obtained and the resources used) and effectiveness (the relationship between the results obtained and the initial objectives).

2.4.4. *Ethics of practices*

– Culture (CUL): The design, implementation and use of data are accompanied by the behavior and habits of the actors in charge of these activities. The culture around data management and the use of an AI is built through awareness, training and education of stakeholders.

– Regulation (REG): The implementation of all the theoretical, material and technical means to maintain each AI equal to a desired value, by action on a regulating AI, despite the influence of the disturbing elements of its lifecycle. Regulation is a specific case of feedback where the device tends to reduce its deviations from the control.

– Deontology (DEO): The set of ethical principles and rules (code and charter of deontology) that manage and guide a professional activity. These standards are those that determine the minimum duties and rights required by professionals in the accomplishment of their act and deed. This professional deontology is directly associated with the principle of personal responsibility. Thus, following the major advances in terms of data science and AI algorithms capable of learning, an awareness is beginning to emerge around the ethical responsibility of developers, publishers and integrators of digital solutions and services. Their profession is particularly solicited on these issues, since they are the ones who supervise machine learning, from databases that may contain biases, which implies an ability to be impartial and to take these issues into account by default and as far upstream as possible. Publishers and integrators have *de facto* strong responsibilities on the way digital solutions or services are designed and arranged.

– Trustworthiness (TRU): Reliability, credibility, non-discriminatory aspect and, therefore, loyalty are essential components for the trust of individuals about the algorithm that uses their personal data. This requires a perfect knowledge and understanding of the various operations that make up the processing algorithm. This notion is naturally linked to that of transparency.

– Privacy (PRI): The ability of a person or group of people to isolate themselves in order to protect their interests. Privacy can sometimes be equated with forgetfulness, anonymity, or a desire to stay out of the public sphere. Privacy is often linked to security aspects and individual freedom (protection and autonomy). The degree of privatization of information is influenced by the way the public environment might receive it, which differs according to space and time. Confidentiality and data protection must be ensured at all stages of the AI application lifecycle. This includes all the data provided by the user as well as all the information produced about users during their interactions with AI. Organizations must be aware of how data are used and how it could affect users, and must ensure full compliance with the General Data Protection Regulations (GDPR) and other applicable privacy and data protection regulations.

– Accountability (ACC): The obligation for companies to put in place internal mechanisms and procedures to demonstrate compliance with data protection rules used by the algorithm. It is the idea of "being accountable", whether voluntarily or otherwise.

– Inclusiveness (INC): AI systems should empower everyone and mobilize people:

- AI systems must take into account and integrate a wide range of human needs and experiences;

- the establishment of diverse design teams can make it possible for developers to understand and anticipate the variety of user needs and thus not exclude individuals;

- AI systems must also understand the context of the decision they have to make and develop their emotional intelligence;

- AI systems can allow people on the margins of society to be better integrated (compensation for disability, etc.).

It is, therefore, necessary to ensure the participation and inclusion of stakeholders in the design and development of an AI. This also requires a diversity of stakeholders during the constitution of development teams, implementation and testing of the product.

2.4.5. *Ethics of decisions*

– Dehumanization (DEH): Data science can result in reducing the person to a data avatar. There is, therefore, a risk of depersonalization of the individual, which has a direct influence on the accountability of those who deal with the data and forget that behind every digital profile is an individual.

– Autonomy (AUT): It defines the ability of a person to govern oneself, according to one's own rules. The individual is able to function independently, without being controlled from the outside. Autonomy is often associated with the free and informed consent of the person to become a full-fledged actor. Therefore, the latter should be informed in clear and understandable language of the binding or non-binding nature of the solutions proposed by AI tools, of the different options available and of one's rights. In addition, the autonomy of the user must be reinforced and not be restricted by the use of AI tools and services. The professional should at all times be able to revert to the decisions and data that were used to generate a result and continue to have the possibility to deviate from them in view of the specificities of the situation. Good governance of AI self-nomination includes, for example, a greater or earlier human intervention depending on the level of social impact of the algorithmic system. It also includes the difficulty by which an AI user, especially in a work or decision-making environment, is allowed to deviate from a process or decision chosen or recommended by AI.

– Free will (FWI): It is the ability of a person to determine freely and by oneself (without outside influence) to act and think. This question arises when we study self-learning expert systems or other decision support platforms and devices in relation to the professionals whose mission is to make that decision.

– Management (MAN): Algorithmic processing management enables the development of architectures, practices and regulations to be considered throughout the data lifecycle. Along with processing, data are one of the two aspects of IS traditionally identified, and both are essential for coherent IS management. Algorithm management is an approach to the management of an algorithmic system that is structured around the nature of the algorithm and its interactions with the environment.

– Governance (GOV): It includes the supervision, steering and regulation of algorithms at the societal level. Many questions are centered around the place that algorithms take in our society and a certain form of "algorithmic governance". Poor governance, by which it becomes possible to intentionally or unintentionally alter data, or to give unauthorized entities access to algorithms, can lead to discrimination, erroneous decisions or even physical harm to the individual.

– Responsibility (RES): Responsibility is the principle that a person held politically or legally responsible for a prejudice has an obligation of justification or compensation in one form or another. However, no one can be held responsible unless he or she has some degree of control, in the sense that he or she facilitated or caused the harm or is able to prevent or mitigate it. From a legal point of view, liability manifests itself through the concept of an obligation to make reparation (e.g. damages). As a general rule, the law attributes liability to the person who is in a position to prevent harm or mitigate a risk (through insurance or otherwise). Accountability for algorithmic decision-making is complicated by the fact that it is often unclear who has the degree of control necessary to be held legally or politically liable. A case in point is the case of algorithmic tool developers, who may not know anything about the future use and implementation of their tools. Conversely, people who integrate algorithmic tools into applications may not be aware of the future use and implementation of their tools. Should the responsibility be attributed to the people who developed and coded the algorithm? Some authors argue that liability and regulation of algorithms is impossible because programmers themselves are unable to predict or fully understand how the algorithm makes its decisions. Another area to explore is whether existing product liability regulation should be extended to include software. Liability schemes can range from monetary compensation (no-fault insurance) to search for the fault and reconciliation without monetary compensation. The choice of liability processes may also depend on the nature and impact of the activity as well as the level of autonomy involved. Are the responsible parties to be sought rather on the side of public and private actors, who buy the algorithm and use it to offer their services without even understanding how they work? Algorithmic responsibility must be reinforced by procedural guarantees and the rule of law. It is also essential that anyone whose rights would be violated by automated decision-making systems should have access to effective redress mechanisms. Individuals who design and launch AI systems must be accountable for the operation of their systems:

- those who develop and implement AI systems must be held accountable for the systems they have developed;

- internal control and audit panels can develop best practices in this regard.

– Announcement (ANN): The disclosure of an information, result, or decision from an NICT is a decisive moment for the person. This announcement must respect and maintain the autonomy of the individual by making him/her share the responsibility of decisions concerning him/her (free and informed consent). This delivery of information needs to be optimal, adapted to the person, consistent over time, evoked and shared, and above all reliable.

– Environment (ENV): It includes the set of elements (biotic or abiotic) surrounding an algorithmic system, some of which contribute directly or indirectly to its needs, or even to its development. This can represent living beings, connected objects, digital applications, etc. It can also be described as the set of natural (physical, chemical, biological) and cultural (sociological) conditions likely to act on NICTs. The notions of ecological, social, racial and sexual diversity are associated with the environment. AI systems should treat all people equitably:

- AI can contribute to better decisions, since computers have a purely logical reasoning in theory;

- it is necessary to train people to understand and interpret the meaning and implications of AI results so that they can make a decision.

Finally, digital technology is generating strong growth in energy consumption, both in terms of usage and equipment production. The digitization of the world (with the increase in data flows, processing, storage, etc.) is producing significant negative effects on the planet (global warming, extraction of rare metals, health impacts, etc.). However, awareness of the energy, carbon and ecological footprint of digital technologies is still low.

The main principles of eco-design are as follows: selection of low-impact materials, reduction of material use, optimization of production techniques, optimization of the distribution system, reduction of impacts during the use phase, optimization of the initial lifecycle and optimization of the end-of-life system.

In conclusion, these various indicators that make up our neo-Darwinian prism of Ethics by Evolution provide a foundation on which can be structured, over time, the framework of algorithmic processing, and thus feed a repository and then an ethical charter around digital projects related to AI.

2.5. Analysis of AI-related knowledge

Having a "complex thought" leading to a philosophy of solidarity and non-coercion seems necessary to reach the ultimate stage of a society called "ethical information". Complex thinking nourishes reflexive ethics[33], which has the mission of bringing "legitimization" to any practice and any norm. The challenge of modernity is, therefore, to rehabilitate the "relational person", through framed networks of cumulative and non-exclusive exchanges of knowledge. To do this, it is necessary to invent, within a controlled ethical framework, new bonds of proximity that rely on NICTs to create a new art of "living together" that is de-compartmentalized. This reflection is guided by ethical values that serve as both normative and critical guidelines. By linking knowledge, it orients toward the reliance between humans and converts knowledge into "practical wisdom" (Ricœur 1990), fundamental in the elaboration of an optimal decision and in which the duty itself must pass the test of wise, prudent decision in the face of singular concrete situations. According to Edgar Morin (2004), its principle of non-separation is oriented toward solidarity. Thus, complex thinking leads toward ethics of responsibility (recognition of the relatively autonomous subject) and solidarity (thinking that connects). It leads to ethics of understanding, that is, ethics of pacification of human relations. It shows that the greater the social complexity, the greater the liberties and the greater the need for solidarity to ensure social cohesion.

Moreover, from our Neoplatonic systemic ethical modeling, we can establish a new representation of knowledge integrating different levels of modeling according to the fields of study. Each field of study corresponds to a concept of the world.

This pyramidal representation tends toward a personalized ethical world assisted by NICTs, which has the characteristics, on the one hand, of increasing the meaning of a decision and, on the other hand, of decreasing the entropy (degree of disorder) of an organization (see Figure 2.3).

This ethical space of analysis takes up the three levels of modeling of classical antiquity in the context of a study of events, namely:

33 Ethics is, therefore, constantly required "to remain alert in its mission of legitimization, brought to mobilize its reflexive capacity with reference to values" (Hoffe 1991).

– the "being given"[34] represented by the epistemological aspect;

– the "sentient world"[35] illustrated by the anthropological field centered on human relationships;

– the "objective reality"[36] characterized by ethical and philosophical thinking. It consists first of all in respecting a deontology and human values so that the virtual integrates harmoniously with reality.

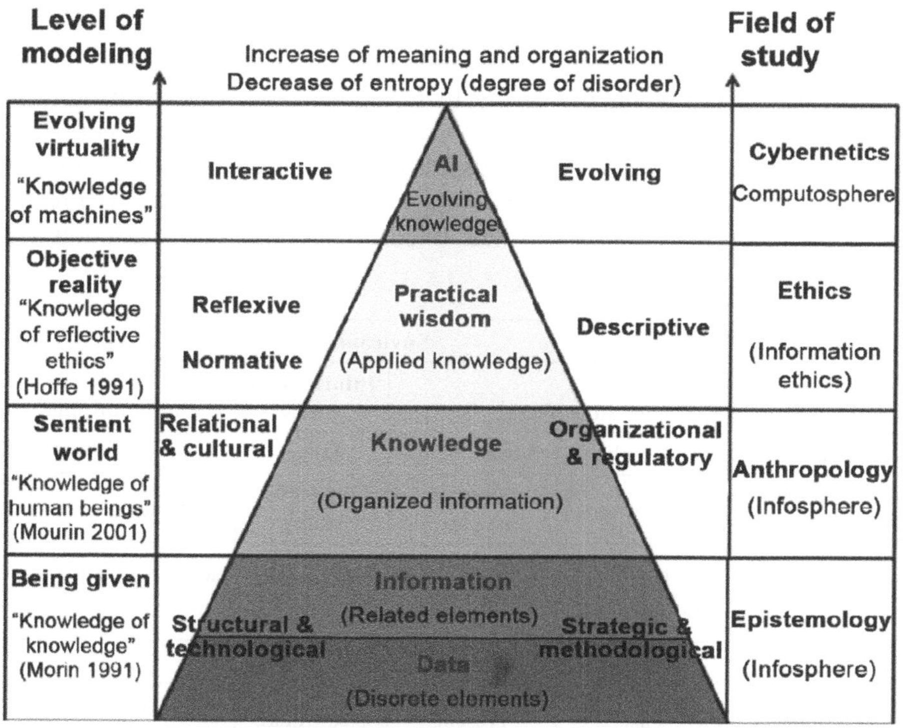

Figure 2.3. *Study of the knowledge pyramid through ethical modeling. For a color version of this figure, see www.iste.co.uk/beranger/responsibility.zip*

34 All that constitutes the world to which humans have indirect or even partial access through their senses.

35 The world of sensation and measurement is the only source of data that provides information about the real world or being given.

36 Space entirely designed by the human mind. It consists of building explanatory models that translate data from the sentient world.

Our modeling was able to highlight the last stage of knowledge processing, which in this case is AI where machines can self-evolve without human intervention. This stage corresponds to the level of modeling called "evolving virtuality" where cybernetics relies on ethics in the computosphere to move toward a world 4.0. This brings us back to the notion of the Internet of Things that Joël de Rosnay has called "4.0", which is linked to the capacity that intelligent machines will have to talk to each other, independently of human will. This plunges us into the universe of Isaac Asimov[37], a science-fiction author, who sought to offer a rational vision of the robot, one of the principles of which is that the machine does not turn against the human being and makes them a possession in its own right. All the resulting problems emanate from its creator. The notion of computosphere is interesting because, while the computer orders and still keeps a link with its designers, the computer itself calculates without having to account to its designers, since it calculates its own evolution (see Table 2.5).

	Environment			
	Infosphere		**Information ethics**	**Computosphere**
Field of study	Epistemology	Anthropology	Ethics	Cybernetics
Level of modeling	Being given	World of senses	Objective reality	Scalable virtuality
Nature of elements knowledge	Data/ Information	Knowledge	Practical wisdom	Artificial Intelligence (AI)
World	World 1.0	World 2.0	World 3.0	World 4.0
Temporality	Past	Present	Near future	Medium-/long-term future
Nature of relationship	Human–machine	Man–machine	Interhuman	Intermachine
Purpose of decision	Doing it right	Doing it right	Doing the right one	Make it evolve

Table 2.5. *Parameters constituting the environment surrounding knowledge*

37 *I, Robot* is a collection of nine science-fiction short stories written by Isaac Asimov, first published in 1950.

CYBERNETICS.–

A cybernetic system can be defined as a set of interacting elements. The interactions between the elements can consist of exchanges of matter, energy or information. These exchanges constitute a communication, to which the elements react by changing their state or by modifying their action. Communication, signal, information and feedback are central notions of cybernetics and of all systems, living organisms, machines or networks of machines. Computers and all the intelligent machines we know today are applications of cybernetics.

COMPUTOSPHERE.–

A set of digitized data that constitute a universe of information and a half-place of communication, linked to the interconnection of computers. It constitutes the support of the informational interconnection between machines.

INFOSPHERE.–

This word was coined by Simmons (1989) to designate an informational environment where information is generated and propagated, especially Big Data. By its very nature, the infosphere constitutes an intangible and immaterial environment, sensitive to the spatiotemporal fluctuations that digital technology generates.

Thus, the computosphere has three informational characteristics: semantics (the meaning of information), syntax (the flow of information) and lexicon (the stock of information). After recalling the positions of Turing and Grice on this subject, the semanticist François Rastier (2001) proposed six "precepts" conditioning an evolved dialogue system, specifying that they were already implemented by existing systems:

– objectivity (use of a knowledge base by the system);

– textuality (taking into account interventions of more than one sentence, whether from the system or the user);

– learning (at least temporary integration of information from the user's comments);

– questioning (request for clarification from the system);

– rectification (suggestion of corrections to the question asked, when necessary);

– explanation (explanation, by the system, of a previous answer).

He also suggests that the system should be able to make itself a representation of the user it is dealing with, to adapt to him or her. The user, on the other hand, tends to adapt to the system as soon as he or she understands that it is addressing a machine, which has the pragmatic advantage for the designer of simplifying certain aspects of the dialogue. As far as cybernetics is concerned, it constitutes the set of theories on control and communication processes and their regulation in living beings, in machines and in sociological and economic systems. Its main object is the study of the interactions between "governing systems" (or control systems) and "governed systems" (or operational systems), governed by feedback processes. Hence, the term "cybernetics" comes from the Greek word *kubernesis*, which figuratively means the action of directing and governing. The organization is active, self-organizing and is dependent on and supportive of the environment. The organization also reacts to information. Information is a stable configuration of a symbol that is both sign and signified. It allows the organization to adapt its behavior at any moment by regulation, transforming and rebalancing itself in order to be in osmosis with the environmental parameters of the infosphere, information ethics and the computosphere.

Therefore, information gives rise to a process of permanent adjustment of the organization through the channels (the system adapts by accommodation) and codes (the system adapts by assimilation) of communication in relation to a project. To represent the organization, he proposes a model consisting of a decision system, an information system and an operating system.

In addition, this pyramid schematization involves control filters for each period of world maturity and field of study (see Table 2.6):

– visualization/perception filter and design/implementation filter in epistemology (world 1.0);

– mapping filter and use in anthropology (world 2.0);

– filter of affectation, regulation and legitimization in ethics (world 3.0);

– evolution filter in cybernetics (world 4.0).

Level of knowledge	Nature	Content	Function	Filter
Data: discrete elements	Epistemology (world 1.0)	Numbers	Categorize	Visualization and perception (what is it?)
		Codes	Calculate	
		Tables	Collect	
		Databases	Measure	
			Collect	
Information: related elements	Epistemology (world 1.0)	Phrases	Contextualize	Design and implementation (how?)
		Paragraphs	Compare	
		Equations	Order	
		Concepts	Converse	
		Ideas	Filter	
		Questions	Frame	
		Simple stories	Prioritize	
Knowledge: organized information	Anthropology (world 2.0)	Chapters	Structuring	Mapping and use (why?)
			Memorize	
		Theories	Understanding	
		Axioms	Interpreting	
		Conceptual frameworks	Evaluate	
		Complex stories	Tear down	
Practical wisdom: applied knowledge	Ethics (world 3.0 – Ethics by Design)	Books	Protect	Implementation, regulation and legitimization (which is better?)
		Paradigms	Embody	
		Systems	Adapt	
		Religions/ beliefs	Synthesize	
		Philosophies	Apply	
		Traditions		
		Principles		
		Truths		
		Schools of thoughts		
Artificial Intelligence (AI): evolving knowledge	Cybernetics (world 4.0 – Ethics by Evolution)	Black box	Develop	Evolution (how to make things evolve?)
		Software analysis	Objectivize	
		Feedback process	Explain	
		Processing algorithm		

Table 2.6. *Structuring of the knowledge pyramid*

Ultimately, complex thinking links epistemology and anthropology in a loop. Epistemology makes it possible to conceive an anthropology, which is a primary condition for philosophical thought, which is integrated into a loop where each other step is necessary for others to reach an ethics. Finally, cybernetics will rely on information ethics to develop AI. This should alert us to the fact that if the 3.0 world is not centered on an ethics of human relations from the conception of NICTs (Ethics by Design), then humans will be dependent on a 4.0 world where our existence and our future will be under the influence of rational and evolving decision of machines (Ethics by Evolution). We must then call upon deontology and human behavior to guarantee the ethics of AI. This necessarily requires an ethical charter surrounding the design, implementation and use of intelligent artificial agents.

3

Ethical Framework Associated with AI

Today, this question of both human and political construction of the algorithm is becoming indispensable because algorithms have penetrated all the domains of our daily lives and structure our access to information. Industry 4.0, based among other things on the algorithmic processing of Big Data, must bring meaning and intelligibility to these data.

> INDUSTRY 4.0.–
>
> The concept of "Industry 4.0" was first introduced at the Hanover Congress in 2011. It represents a new way of organizing the means of production based on technological foundations such as the Internet of Things, Big Data technologies and cyber-physical systems. The aim is to set up so-called "smart factories" capable of greater adaptability in production and more efficient use of the resources at their disposal.

This necessarily involves semantic and ethical algorithms, as well as the construction of ontologies that consist of representing the stakes and risks, and teaching machines to work for the good of the human species and its ecosystem, tending toward information ethics. It is necessary to claim that the human dimension associated with ethical values must never be underestimated or forgotten by the potential offered by NICTs. Ethics is a bridge between the innate and the acquired for our social behaviors and our intelligence. It is the eternal confrontation between the innate and the acquired – in other words, between the potential at birth and the influence of our experiences resulting from our interactions with the external environment. If we take the example of health care, we believe that we will have to wait until 2030 before algorithmic applications in health care are sufficiently standardized, robust and reliable to be accepted by the majority of regulatory bodies, public institutions, professionals and healthcare users. It should not be forgotten that the relationship of trust between doctor and patient is still strongly rooted in our societies, playing a predominant role in the patient's recovery. Many people still prefer to provide their personal data to their doctor rather than to a technological entity. In any case, the medical profession is gradually evolving with the use of technologies that measure the patient's health status on a daily basis in order to make medical decisions. Therefore, the development and implementation of these ethical algorithms may be complex and may cause not only latency, but also bottlenecks in the data value chain. In order to overcome and resolve these complications, it is crucial to identify potential blocking areas, study them and then implement a solution to make the digital data processing system more fluid. It would be interesting to involve healthcare players (particularly healthcare professionals and users) in the design and development of processes capable of enhancing and improving Big Data to support an optimized healthcare journey. To do this, we can take the case of Living Labs, true environments of innovation and participatory design. This ethical use of digital technology in healthcare is a fundamental prerequisite for the development of Medicine 4.0, so that technological creation tends toward nature and human benevolence. Of course, the reasoning that we apply in the medical sector can very well be transferred and transcribed to all other fields of human activity.

> LIVING LABS.–
>
> The Living Lab is a methodology where citizens, inhabitants and users are considered key actors in the research and innovation processes. The aim is to promote open innovation, share networks and involve users from the very beginning of the design process.

3.1. Ethical charter around AI

Society can no longer ignore the constant changes brought about by NICTs in the behavior of each of us, in our professional and personal relationships. Indeed, the evolution of information technologies is having considerable repercussions at all levels: it is transforming society as a whole, but also organizations and institutions. It also changes all social and human interactions. New challenges as well as conflicts have emerged. They influence its users, condition them and modify their vision of the world. An information system is mainly useful through the relationships it creates; information is the symbol, the key and the condition of human interaction because it facilitates exchanges. However, information can be a nuisance if it represents a vector of the will to dominate, and becomes an obstacle against transparency. The challenge is, therefore, to create conditions conducive to a healthy interaction between ethical values, political and legal moral standards, industrial strategies, and the protection of patients against possible deviations in the use of their own information via new technologies.

Thus, the good use of AI requires a maturity, a know-how of living, the reverse of the feverishness of some neophytes. The usage culture must, therefore, support the implementation of these tools. These rules of common sense in the service of the ethics of care are too often discarded in favor of the technicality of NICTs. This is why these two worlds that everything opposes – ethics and IT – must learn to coexist together. As a result, companies must launch concrete management methods and means, accessible at all levels of the organization, adapted to the decision-making level in order to move toward overall performance. This enables AI to be implemented and operated in an optimal way, as the basis of the edifice, the reflections and the dialogue of decision-makers at the different levels of management.

In the light of these reflections, it seems essential to bring to the heart of the implementation and operation of AI both sociology and ethics in charge to produce a conceptual framework of good practice for these data. A good use of personal data based on organizational intelligence necessitates the elimination, prioritization and sorting of accessible data in order to give them meaning and coherence, not to accumulate them. This means that a technology without ethics and knowledge is equivalent to a body without soul and spirit. Under these conditions, it seems to us essential to establish an ethical charter that takes up all the problems of this context in perpetual evolution. This is why it is necessary to set rules, standards and governance norms in order to build an ethical charter on algorithms and autonomous intelligent systems.

Therefore, this ethical charter and these recommendations should not be used as a rigid framework, but rather as a flexible structure to be integrated into the design, implementation and use of AI. Such a document will set a number of limits, forbidding the development of technologies that may affect human rights, dignity and even human life. This charter is addressed to public and/or private actors in charge of the design and launch of artificial intelligence tools and services based in particular on the processing of cloud data that leads to decisions (machine learning or any other method resulting from data sciences). It also concerns public decision-makers and/or private companies in charge of the regulatory and statutory framework, the development, audit or use of such tools and services. Indeed, whether they are designed to provide support for consultation, drafting or decision support, or strategic orientation, it is essential that the algorithmic processing is carried out under conditions of transparency, neutrality and loyalty certified by independent expertise external to the operator.

In addition, the ethics charter promotes the construction of an environmental ethic by introducing fundamental ethical principles that apply to the organization as well as to managers and employees. Such an ethical charter would allow manufacturers to commit themselves, if they so wish, to respect certain ethical and moral requirements regarding the protection of personal data in order to demonstrate and guarantee to their customers their trust and commitment in this domain. In this context, this code of good practice should outline a new framework for the confidentiality of digital personal data, focusing more on the accountability of the use of data by the organizations that collect, cross-reference, analyze and exploit them, and less on individual consent at the time of acquisition. Such accountability could mean that companies (data aggregators) should provide access and rectification rights to ordinary health users (data owners and end users) so that they have better control over the cross-referencing and retention of their personal data. However, it can be noted that the transition from "privacy by prior consent" to "privacy by accountability" will not be possible without significant ethical awareness and education among these actors. Therefore, we propose the proclamation of an ethical charter on the realization and use of responsible AI with a "human face", which aims to make the rights and fundamental ethical principles applicable and adapted to the current context of the world, while placing the citizen and professionals at the center of the debate. This charter must, therefore, focus on the moral sense of individual behavior toward oneself and one's entourage in a professional context. To do this, this document will be based on the 36 ethical criteria divided into five families of digital ethics that we saw earlier.

This charter aims to give all actors involved in AI the ethical, legal and technical foundations useful and necessary for the design, implementation and correct use. It serves as a reference and behavioral guide for individual decisions taken in good faith. It avoids the pitfall of hypertrophied regulations to lengthen in the face of intrusion of AI, by appealing mainly to the professional conscience of each person, knowledge of one's profession and protecting the individual freedoms of all users. This charter intervenes according to two levels of action:

– as a reminder of the overall ethical principles that should normally govern the practice of a profession in order to avoid imprudence, lack of foresight or negligence;

– as a reference to be referred to in order to clarify professional issues and duties.

This is why this ethical charter must be mastered, understood, integrated and applied (as far as possible) by all digital players. The ethical, legal and technical foundations, thus, reinforce the mission of companies that design and/or use AI by developing the best professional practices while respecting the rights and freedoms of all. Before presenting and explaining this ethical charter, it is useful to make the following preliminaries around this notion:

– the definition of rights, values and principles implies that citizens and care providers assume their responsibilities. Social and human concepts are not without responsibilities and duties;

– the charter applies to AI manufacturers, politicians, digital managers and users;

– the charter intends to pronounce on the ethical questions raised by the technical aspect of AI;

– the charter defines rights, values and principles that are valid in the current state of the French healthcare system. It will, therefore, have to be reviewed and modified in order to adapt to its evolution as well as to the development of scientific and technological knowledge on the subject under discussion;

– ethical principles are part of social and human values and, as such, must be recognized and respected independently of financial or economic constraints;

– compliance with these ethical principles implies the fulfillment of technical, logistical and organizational requirements as well as requirements for professional behavior. The social values associated with these principles, therefore, require a global reform of the way AI is operated and used with the application of standards applied to their design;

– each article of the ethical charter includes a definition and non-exhaustive specifications;

– this ethical charter has an experimental value that will be invoked over time according to the development of technologies that will vary and modify ethical problems. This charter is, therefore, not set in stone and must be adapted according to the events with which the actors will be confronted.

Moreover, this document is original in that it gives equal importance to events, whether scientific, technical or contextual, and to the ethical references and norms derived from pertinent guidelines, principles and theories. From this point of view, disciplinary knowledge is essential to the process, since it provides the expertise needed to evaluate the problem in question, to define possible interventions, as well as the means developed by the actors. As conceived here, this ethical charter is built in partnership in the elaboration of solutions that will reach a consensus on the goals, achievements and means concerning AI, while respecting the ethical principles and standards specified according to the situation and the socioeconomic context in question.

In addition, it can be noted that the role and missions of an ethics committee are mainly to:

– identify the norms and rules associated with the human values of a community associated with AI;

– implement and integrate the standards, rules and values of this community within AI;

– evaluate the alignment and compatibility of these standards, rules and values in the interfaces and human-machine dualities of this community.

Therefore, this ethical charter around AI digital projects is composed of 30 ethical commitments, associated with the 36 ethical criteria of an algorithmic processing mentioned before. These recommendations can be classified in an equivalent way according to the families of digital ethics seen previously (see Table 3.1).

Note that this charter is based on the decisive and implicit assumption that designers, developers and users of AI comply with fundamental rights and all regulations in force. The respect of this ethical charter does not replace the respect of the latter, but is simply an operational complement.

3.1.1. *Data ethics*

Data ethics are as follows:

1) Establish traceability of access to digital data associated with AI (traceability).

2) Ensure the quality, integrity, source and choice of the right information transmitted to the person, in accordance with the best standards (integrity).

3) Establish a duty of security and protection of digital data (security).

4) Respect the same rule of access and dissemination of information according to the profile or status of the person (accessibility).

5) Respect the rules of collection, storage, hosting and distribution established by the legislative regulations in force (non-selective collection).

6) Match the use of numerical data with the organization of the structure that uses AI (organization).

3.1.2. *Ethics of systems*

Ethics of systems are as follows:

1) Provide citizens with ergonomic simulation interfaces so that they can appropriate the functioning and effects of the algorithm (ergonomics).

2) Adapt and make AI applicable to the knowledge and know-how of professionals (adaptability, applicability).

3) Ensure the technical and scientific relevance of the tool's automation (automation).

4) Develop an AI oriented toward collective performance (performance).

5) Establish consistency between the functionalities of AI and the work organization initially set up (coherence).

6) Integrate parameters and confidentiality measures around the architecture and structure of the AI (confidentiality).

3.1.3. *Ethics of algorithms*

Ethics of algorithms are as follows:

1) Guarantee the diffusion of an adapted visibility of algorithms, the communication on the role of algorithms and on the description of algorithmic operations, as well as information on the existing recourses ensuring the effectiveness of the rights of individuals (transparency, explainability).

2) Work for the good and the social benefit of the individual, or even of the community and humanity (purpose).

3) Ensure the reliability and quality of algorithmic operations (reliability, quality).

4) Reduce unnecessary or miscalculated risks as well as unfair biases and discriminations against illegal or illegitimate criteria, either directly or indirectly (bias).

5) Implement a protection policy throughout the operational life of AI through regular testing and evaluation of the algorithmic system (protection).

6) Verify that the increasing and evolving empowerment of AI always serves the interest of the individual or even the general interest of society (empowerment).

3.1.4. *Ethics of practices*

Ethics of practices are as follows:

1) To set up internal mechanisms and procedures to show the loyalty of the algorithm by adhering to the rules relating to the protection of the data used (accountability, trustworthiness).

2) Take into account and integrate a wide range of human needs and experiences in all their dimensions (inclusion).

3) Ensure equal and/or fair treatment and the guarantee of respect for the fundamental rights and freedoms of all citizens (deontology).

4) Ensure that the respect and protection of privacy (Privacy by Design) of individuals (whose data are used for any use or algorithmic prediction) are taken into account from the design of AI tools (privacy).

5) Reinforce the transversality of the people involved in the AI digital project (culture).

6) Establish a policy of management and steering on the use of AI, considering the environmental, societal and economic impact (regulation).

3.1.5. *Ethics of decisions*

1) Establish individual responsibility – within the limits of their level of information, decision and action – for those responsible for the design, use or control of algorithms (accountability).

2) Obtain the free and informed consent and adherence of the individuals or institutions from which the digital data that feeds AI is derived (autonomy).

3) Integrate within the goals and behavior of AI the human values based on human rights, freedoms, human dignity, social and cultural pluralism, diversity, and respect for the societal and ecological environment (environment).

4) Support the decision established by AI of a satisfactory explanation, even of a control, by the human being (announcement, free will).

5) Ensure that the launch of the AI algorithm on data imperatively respects the purpose for which access to the data has been obtained (governance, management).

6) Put human beings back at the center of decision-making by bringing them more complete, understandable and quicker information from AI (dehumanization).

> PRIVACY BY DESIGN.–
>
> Privacy by Design is one of the concepts at the heart of the GDPR, the new regulation governing data protection. In short, it is a concept that requires companies to integrate the principles of the GDPR from the design stage of a project, service, or any other tool related to the handling of personal data. The idea is to impose that every new technology designed to process personal data must be designed to provide a high level of data protection.

Finally, we can say that the ethical charter represents a tool for regulating the use of AI that enables:

– deterring deviancy and illicit behavior, and raising awareness of digital data security;

– improving the efficiency of AI and Big Data use by users;

– formalizing and enunciating "bilateral" ethical principles around the use of AI, particularly related to issues of electronic surveillance, use of resources for personal purposes and protection of privacy.

In this case, this guide to good ethical practices establishes a kind of moral contract between employees and the employer. Therefore, it constitutes a foundation and a decisive framework for raising the awareness of all the company's staff about their responsibility with regard to their use of the NICTs available to them. It encourages the appropriation of standards and the acquisition of appropriate behaviors. The development of and adherence to this charter also increases citizen confidence in AI actors. Thus, each employee will know in a clear and precise way the obligations he/she has toward the employer, the organization or the company, and finally who is responsible in the case of an ethical problem.

3.2. Recommendations for AI

The development of AI requires us to invent new knowledge, new know-how and new ways of living with these new technologies. It is from this moment that we can glimpse a new era that serenely integrates AI and not an era that gives up, struggles or runs behind. The ethical recommendations relating to digital AI projects are composed of 50 ethical actions associated with the 36 ethical criteria of an algorithmic processing stated earlier. This list of recommendations can be divided equally from the families of algorithmic ethics seen before.

Some recommendations can be undertaken directly by national or European decision-makers, in collaboration with stakeholders where appropriate. For others, decision-makers can play a leading role, a role for efforts undertaken or led by third parties.

We believe that to create a good AI society, the ethical principles identified in Chapter 2 must be integrated into the default practices of AI. In particular, AI should be designed and developed to reduce inequalities and enhance social empowerment, with respect for human autonomy and increase equitably shared benefits. It is in particular important that AI be explainable, as explainability is an essential tool for gaining public trust and understanding of the technology.

We also believe that building a good AI company requires a multi-stakeholder approach, which is the most effective way to ensure that AI meets the needs of society by enabling developers, users and decision-makers to participate and collaborate from the start.

In the European Union (EU), the protection of personal data is currently governed by Regulation 2016/679 of April 27, 2016, and the High-Level Expert Group for AI emphasizes that this protection must be guaranteed throughout the lifecycle of an AI system, so that it applies equally to personal data. The only personal data that can be entered into the system are the personal data generated by the system after processing of the initial data. This is perfectly in line with our Ethics by Evolution.

To this end, the idea of integrating a mechanism for obtaining consent from an individual and a mechanism for revoking that consent within AI systems is being discussed. In addition, AI system designers are advised to determine whether personal data are present and, if so, to take measures to enhance the confidentiality of such data (e.g. by strengthening encryption or anonymization techniques for personal data). They should also make every effort to use only as much personal or sensitive data as is strictly necessary for the operation of the AI system they design. Finally, it is recommended that they develop data access protocols that would define who has access to the data and under what circumstances.

Different cultural frameworks inform attitudes toward NICTs. This book represents a European approach that is complementary to other approaches. We commit ourselves to develop an AI in the service of the public interest in order to strengthen a shared social responsibility and thus gain the trust of citizens.

Finally, this set of recommendations should be considered as an "evolving document".

The following recommendations provide an initial outline of actions that could be undertaken. The various points, actions and recommendations are designed to be dynamic, not simply requiring single policies or isolated investments, but rather ongoing efforts to ensure that their effects are sustained and lasting.

3.2.1. *Data ethics*

Data ethics are as follows:

1) Promote equality and equity of access and opportunity (social justice). Voice-based NICTs and social robots offer new accessible solutions, especially for people with disabilities such as visual impairment, dyslexia and reduced mobility. This justiciability can also translate into Open Science (encouraging open research), by encouraging good practices of openness and sharing. AI must benefit and authorize as many people as possible (accessibility).

2) Do not feed AI with numerical data not relevant to its final purpose (integrity).

3) Guarantee the reliability and quality of data ("data repository" approaches). For this, there should be access to the intrinsic value of the data, which could take the form of a data value index, such as a stock market product (integrity).

4) Make an inventory of all the risks of construction and functional bias, discrimination and exclusion that an AI can generate and cause via its data (integrity).

5) Establish a constructive and collaborative exchange between scientific researchers, developers and decision-makers in AI. For this, it is necessary to introduce from the time of design of AI a representation of the stakeholders in the concerned environment to ensure that the concerns and opportunities specific to the sector are well understood and addressed (organization).

6) Adopt a transversal and global vision of AI, by not focusing solely on science or technology, but by mixing from the outset all the actors in order to create a strong territorial added value based on a diversity of viewpoints: ethics, economy, law, science, technology, society, etc. (organization).

7) Encourage the development of an AI powered by Open Data in order to encourage open publication, social impact reporting and the development of data of ecosystem interest (non-selective collection).

8) Analyze the structure of a source, and define the modes of exploitation of these data according to their purpose (business) or their potential[1], in order to use them well (traceability).

9) Set up a device that measures the personalization of the data at the end of the treatment (traceability).

10) Write internal procedures and good practice guides on information protection and security. This can result in the implementation of a security policy throughout the operational life of AI, with verifiable and applicable characteristics and indicators (security).

3.2.2. *Ethics of systems*

Ethics of systems are as follows:

1) Set up an AI with a beneficial and useful purpose for the person and/or the society. AI must maximize efficiency without destroying people's

1 This potential is complex to understand, since generally the value is revealed by coupling several types of information together. It is, therefore, essential to evaluate this potential with regard to the other sources available.

dignity: AI must preserve our cultural commitments and foster diversity in order to enhance positive social change and strengthen sustainability and ecological responsibility. It is not for the digital industry to dictate our values. Therefore, it is of paramount importance to develop a responsible AI with strong ethical principles and guidelines (purpose).

2) Ensure that digital applications and solutions have not been designed so as to voluntarily manipulate the user by exploiting cognitive biases (purpose).

3) Do not violate human dignity and respect for the person. To this end, AI must be constructed and operated in a manner consistent with ethical ideals and principles, moral and social values, rights, economics, freedom and cultural diversity. The purpose must be directed toward the welfare and benefit of humanity. For this, it is essential to integrate ethical considerations and recommendations[2] (purpose, coherence) from the conception of NICTs.

4) Preserve the confidentiality of personal data. Thus, AI must not be able to keep or disclose confidential information without the explicit approval of the source of the information. Privacy is contextual and situational. AI must be designed to implement sophisticated protections that secure personal information and information about groups of people in such a way that trust in these technologies is justified (confidentiality).

5) Work the design of AI in the service of human freedom to counteract the "black box" effect. AI must increase human capabilities, not replace them. This also applies to the labor market. It is essential to design technologies that complement rather than replace human work, while encouraging business leaders to change our culture, often focused on labor savings and automation, toward a mindset more oriented toward manufacturing and creation (ergonomics).

6) Integrate playability within an AI in order to develop interaction to understand, improve and make it responsive. To do this, the AI application must consider all human skills, abilities and requirements to ensure its accessibility (applicability).

2 More specifically, ethics will have to be adhered to at the data collection stage (loyalty), then at the heart of the algorithm, as a value for managing data throughout its lifecycle, and finally in AI practices by explaining the results produced in adequacy with the purpose of the collection.

7) Establish a methodological monitoring tool that makes it possible to reference existing and future legal and regulatory obligations (applicability).

8) Implement feedback to improve AI. It is, therefore, necessary to integrate feedback, a virtuous feedback loop, improve and make devices robust (automation).

9) Reward AI for good results and good decisions. For this, reinforcement learning methods must, on the one hand, identify what AI must do in order to achieve a given result or decision, and, on the other hand, ensure that these actions are associated with moral and human values. Therefore, AI must be carried out in such a way that its objectives and behaviors are consistent with initially predefined human values (performance). For this, it is essential to establish, on the one hand, a study of the individual and collective interest and benefit of an algorithmic autonomous system, and, on the other hand, an analysis of AI performance compared to human action (sandbox method).

10) Train and raise awareness of all actors (designers, professionals and citizens) of the algorithmic system and the digital data development cycle to ethics. Digital literacy must enable every human to understand the springs of the machine (adaptability).

3.2.3. *Ethics of algorithms*

Ethics of algorithms are as follows:

1) Make the purpose of AI fully transparent to its users and society. Human beings must be aware of how the technology works and what the rules are. We want machines that are both "intelligent" and "intelligible". To know how and why an autonomous system makes its decision, on the basis of which parameters. Not artificial intelligence, but symbiotic intelligence. Technology will know things about humans, but humans must also know about machines. Citizens should have a deep understanding of how technology sees and analyzes the world. This transparency and clarity of AI processes can be achieved through certification and/or control audit, or even investigation (transparency). One could imagine depositing the source codes of algorithms within an agency for the protection of programs (APP).

2) Design a technological solution (declarative and/or explanatory button, black box system, etc.) that improves and simplifies the transparency of AI. This will allow processes, related data flows, performances, limits and risks of AI to be documented[3] (transparency). For this, an analysis on biases must be performed, especially before the constitution of the AI learning or testing sets. This risk study must be revised throughout the process. It would then be useful to provide a mapping of the main risks identified for this assessment.

3) Impose on AI to clearly identify itself as a technological system, without being able to pretend to be human (transparency).

4) Regularly test and evaluate the efficiency, usefulness, security and robustness of AI so that it can handle errors and/or inconsistencies during all stages of the lifecycle of AI applications. To do this, it is necessary to interact, play with parameters, and let people improve the latter, in order to allow mediation and explainability. It is necessary to consider developing new standards and rules that describe the measurement parameters, the degrees of transparency of autonomous systems, so that these systems can be objectively evaluated with determined levels of compliance (reliability).

5) Give AI designers the opportunity to be able to determine the cause of the damage(s), in case of AI malfunction. This requires the establishment of recording systems (by the designer and/or user) of autonomous systems that include intended uses, input and output data, sensors, data sources, processing algorithms, graphs, characteristics of the model in place, user interfaces, optimization objectives/loss and reward function (explainability).

6) Anticipate the risks likely to be caused by AI; moral, catastrophic or existential risks must be predicted in order to limit their impact (protection).

7) Develop algorithmic development practices that incorporate verification processes against discrimination, bias and racial, ethnic, religious, sexual prejudices, etc., but also incorporate positive devices for the protection of fundamental rights, consumer and citizen rights (bias).

8) Train the designers of the algorithm to the risks of bias related to the data sets used for machine learning (bias).

---

3 The criteria for such documentation could be verifiability, accessibility, meaning and readability.

9) Give the possibility to AI to be able to erase or delete data and/or results at the request of its designer (empowerment).

10) Ensure that AI research and technology is robust, reliable and functional within secure (quality) limits.

3.2.4. *Ethics of practices*

Ethics of practices are as follows:

1) Establish a culture of cooperation, prevention, trust, openness and transparency between the different actors involved in AI. AI developers, scientists and decision-makers must work in a cross-cutting and coordinated way to avoid shortcuts with safety rules and standards (culture).

2) Cultivate the idea of a “popular AI” by sensitizing and educating as many people as possible so that everyone can easily develop their AI tools for their own needs (culture).

3) Ensure education and safety in the field of ethics applied to digital technology that raises awareness of potential risks of AI misuse (culture).

4) Popularize and make AI accessible to all, by implementing new playful and interactive ways to educate and raise awareness among a large number of people about the challenges and risks of autonomous systems (via social networks, YouTube, comic books, training workshops via serious games, etc.) (culture).

5) Establish a gradation of the regulation associated with AI according to the level of sensitivity of the data and the nature of the purposes (regulation).

6) Anticipate, with the help of teams specialized in foresight and strategy, the impacts (environmental, societal and economic) of technological transformations for the company’s businesses and activities. We can set up processes of insurance and investigation in the case of an accident or injury of AI on humans (regulation).

7) Design an AI system that is auditable, supervised and open to inspection by regulatory and standards bodies, throughout the operating

lifecycle of an AI (security, traceability and transparency). It is essential that AI be verifiable, evaluable and documented in order to allow for its control, comprehension, replicability and to be able to prove how it works (accountability).

8) Understand and respect the interests of all parties that may be influenced by AI advances (trustworthiness, inclusion and diversity).

9) Develop a code of good conduct for the general organization around AI. This requires developing rules and involving all the parties directly or indirectly concerned with the algorithmic system throughout its lifecycle. The objective is to involve them and to make them appropriate. A process of participation in the development of AI must be put in place, including the involvement of citizens (inclusion and diversity).

10) Ensure that AI must not be used to diminish the rights and/or privacy of individuals (deontology, privacy).

3.2.5. *Ethics of decisions*

Ethics of decisions are as follows:

1) Reflect the diversity of its users and society as a whole. It is essential, on the one hand, to consider during the design and development of an AI, the diversity of existing cultural and social norms of users, and, on the other hand, to set up a multidisciplinary ecosystem that will make it possible to establish recommendations, standards, rules and limits around AI, and their social impacts. To do this, special attention must be paid to the training of developers and especially at the level of its recruitment (gender parity and social diversity) in order to avoid bias around the learning of AI that is too simplistic, reproducing stereotypes and actions representing only a small part of society. This diversity must make it possible to effectively filter prejudices and negative feelings within the digital data that feed AI. Hence the need to diversify the origin of developers, of AI coders, so that they better reflect the plurality of human values, so as not to reinforce divisions in society, and not to favor weapons of radicalization, but also to push "non-technical" profiles to be interested in and understand the code (environment).

2) Think about the human-machine-environment triptych rather than just the human-machine relationship in order to conduct a regular assessment of

the AI's environmental footprint (at most every 2 years), based on recognized and "auditable" indicators (Green IT or WWF France). The objective is to develop AI for the benefit of humanity as well as to serve the planet. Indeed, within NICTs that emit a large amount of heat, a system can be installed that enables this surplus heat to be reused for ecological purposes to avoid energy waste (digital ecology). For this, it will be necessary to set up research that considers the stakes and questions around the digital environment (environment).

3) Make AI as well as its users accountable. AI must have an algorithmic responsibility so that human beings can cancel any unintentional attack. We must design these technologies not only for the expected, but also for the unexpected. It is necessary to establish a climate of trust that inevitably implies responsibilities on the part of the actors. This self-learning technology must be accountable for its actions and decisions, just as human beings are. This responsibility can be achieved through mediation by organizing dialogue and exchange between systems and society. For this, it appears necessary to identify the team or person responsible for AI (Chief Algorithm Officer) (sandbox method[4]). In addition, the designers and builders of AI systems must be responsible for the moral consequences resulting from their misuse and their actions. For this, it is important to clarify, identify and prioritize the issues of culpability and liability related to AI. In a context of globalization, interaction of value chains and economic flows, the question of liability will have to be studied at the international level in order to anticipate, for example, differences in the appreciation of jurisdictions in the case of possible disputes around an AI system (e.g. components developed or manufactured elsewhere than where the system was used) (liability).

4) Do not allow AI to unduly restrict the individual freedom and the real or perceived privacy of the persons concerned by the application of AI (autonomy).

5) Retain the human power to decide which decisions to make, exercising their freedom to choose where appropriate and cede it in cases where

4 On the ethical, deontological and regulatory levels, the "sandbox" method could clarify the responsibilities involved in case of error and map the multiple hypotheses of legal risks incurred by the use of AI. The opening of the "sandbox" has more players (i.e. European or international players) and will be all the more beneficial for a global and not strictly national approach.

compelling reasons, such as efficiency, may outweigh the loss of control over decision-making. As expected, any delegation should remain voidable in principle (decide to take a new decision) (autonomy).

6) Establish specific governance and external auditability protocols for AI (governance).

7) Reconnect engineers with the humanities and social sciences (HSS). Developers need to work more with psychologists, sociologists and cognitive science experts. It is essential to work in a multidisciplinary, diversified and cross-discipline way and not in silos and in isolation[5] (management).

8) Integrate human control (human guarantee)[6] throughout the AI lifecycle. The designer and/or user must be able to choose whether or not they wish to delegate tasks to AI in order to achieve their objectives. In addition, it is crucial that the user and the decision-maker of AI be able to parameterize the decision-making or results of AI according to their vision and knowledge of the field. Finally, the self-development or self-replication of AI must be subject to regular and rigorous security control (dehumanization).

9) Create awareness and training mechanisms to provide keys to understanding and decision-making that are accessible to all (free will).

10) Warn about the capabilities and potentialities of AI. In the case of a non-consensus, the AI designer must avoid making strong assumptions about the upper limits of the abilities and skills of AI (announcement).

---

5 In antiquity, disciplines were not as compartmentalized as they are today. One was at the same time a biologist, a doctor, a philosopher, an astrologer, a mathematician and so on. For example, Galileo said that he practiced "natural philosophy". Newton's book is called *The Principia: Mathematical Principles of Natural Philosophy*. Then, with the advent of encyclopedists, especially when Diderot was obliged to categorize disciplines, specializations appeared.

6 The principle of the "human guarantee" of AI (human oversight or human warranty) has already been the subject of multiple projects in the framework of the current process of revision of the French bioethics law under the aegis of the CCNE. In the case of AI, the idea is to apply the principles of regulation of AI upstream and downstream of the algorithmic system itself by establishing points of human supervision. This should be done throughout the lifecycle at critical points identified and selected in a shared dialogue between professionals, citizens and innovation designers.

OPEN DATA.–

It is the fruit of its time, when the imperatives of transparency and accountability are more and more important. Transparency is seen here as the answer to an era of mistrust, or even distrust, of institutions and their representatives. This movement responds to a range of economic and political issues. From the opening up of data, we expect democratic benefits (better transparency of public action, citizen participation, response to the crisis of confidence in elected officials and institutions), but also the creation of economic value through the development of new activities based on open data.

SANDBOX.–

The "sandbox" is used to test the potential of AI, particularly in the healthcare sector, for diagnosis and treatment as well as for clinical trials, drug traceability, healthcare reimbursement and the financing of e-health projects.

To these ethical recommendations, we can associate the following recommendations that emanate from the preliminary work for the elaboration of Occitanie Data's ethical charter, carried out by Anthéa Sérafin during her thesis[7] (2018–2019) for her European Master of Laws degree:

– the collection, the method of processing and the purpose of the processing of the Big Data must respect sustainable development and the fundamental rights of individuals, according to a logic of beneficence;

– projects based on the exploitation of Big Data must be subject to an impact study based on a precautionary approach;

– citizens need to be informed about the capabilities and limitations of Big Data-based economic models, especially in order to be involved in the data governance process (alongside experts and public actors);

– Big Data must be of an optimal reliability (veracity of Big Data);

7 The paper is entitled *Intelligence artificielle et santé: stratégies et politiques de l'Union européenne (UE) et de l'Organisation des Nations unies pour la science, la culture et l'éducation* (UNESCO).

– an independent audit of projects based on the exploitation of Big Data should be carried out on a regular basis;

– the damages resulting from the operation of the Big Data must be able to be compensated;

– the multidisciplinary approach must be favored for the development of projects based on the exploitation of Big Data.

In addition, while AI incorporates new features that facilitate data access, movement, transformation and analysis, it also revisits the following new needs, recommendations and requirements:

– data integration tools must work both on-site and in cloud computing models;

– it is essential to define and depict a diagram on the data accessibility processes via NoSQL and analytical databases, or by Hadoop;

– processing tools must be able to simplify the development and implementation of transformations of unstructured raw data into relevant and consistent information. These changes must be reused and exchanged;

– as the data supply chain becomes more complex, it becomes essential to synchronize data between repositories;

– AI associated with Big Data must contain a flexible structure in order to facilitate a fast and precise exploration;

– it is essential that these technologies incorporate transfer devices that can handle new volumes. Information from the study of huge volumes of data can be sent to applications so that more accurate models of reality can be made. For example, data can be synchronized using an in-memory analytics tool, rather than being reduced to SQL databases;

– a good evaluation of integration technologies necessarily involves present data and Big Data that can be easily integrated and stored in canonical form. It is essential to build canonical forms of different types of information from smartphone applications, social networks, web blogs, etc. The control of modifications made to the classical forms of data is facilitated by a good management of the canonical identifications of the data in a lifecycle;

– AI must remain ergonomic and simple to use in order to facilitate its access to a large number of people who can be directly connected to the data, thus promoting self-created solutions and discoveries;

– AI must support processing at all stages of the data supply chain, and provide automated model discovery and visualization;

– it is fundamental that the mechanisms for integrating Big Data be established to articulate with several types of environments;

– optimal AI should allow for a guided experience in which an automatic learning process makes suggestions before being steered in the right direction by the analysts.

> NoSQL.–
>
> In computing and databases, NoSQL is a family of database management systems (DBMS) that departs from the classic paradigm of relational databases.

> Hadoop.–
>
> Hadoop is a free and open source framework written in Java to facilitate the creation of distributed and scalable applications that allow applications to work with thousands of nodes and petabytes of data.

Furthermore, how can we identify whether a data item is "ethical" and trustworthy?

In the white paper, Business & Decision, published in December 2019 and entitled *Data Éthique – IA Éthique: les deux visages d'un futur responsable,* the ten main characteristics are identified that can enable companies to consider any data or set of data as ethical and reliable (see Box 3.1).

The data must be precisely defined in relation to its ontology[8], i.e. within the framework of the business you wish to model.

It must be referenced in an accessible and up-to-date data dictionary, including its name, type, characteristics and exact definition.

8 In computer science and information science, an ontology is the structured set of terms and concepts representing the meaning of a field of information, be it the metadata of a namespace or the elements of a field of knowledge. The ontology itself constitutes a data model representing a set of concepts in a domain, as well as the relationships between these concepts. It is used to reason about the objects of the domain concerned.

It must be right. In the case of uncertainty, it must be known and recorded with the data.

Its exact date and time of collection must be known and recorded.

Its mode of collection, integrating the different possible sources for these data (for example, a questionnaire collected by telephone and Internet), must also be filled in.

The data must be present or explicitly declared as missing.

It must be consistent, i.e. it varies within the limits defined in the dictionary. Likewise, it must not contradict another value related to the same observation.

It must be unique, i.e. an observation must not give rise to two entries in the same entity.

It must be compliant, legal and validated, i.e. it must comply with internal governance regulations and standards as well as external regulations in force (DPGR).

It must be useful and valuable: data should not be stored in an IS without at least one objective in terms of its intended use or value.

Box 3.1. *Ten key points of ethical data*

GDPR.–

The acronym GDPR stands for General Data Protection Regulation. This regulation, of the European Parliament and the Council of the European Union (EU Regulation 2016/679), was adopted in 2016 for entry into force in May 2018. It aims to harmonize the governance of personal information within the member countries of the European Union, in particular with regard to the security and protection of personal data held by companies.

As soon as a data item does not satisfy one of these 10 points, it becomes less and less ethical. However, it is of course extremely difficult for 100% of the data to be in line with all these characteristics. Instead, organizations must follow an "ethical gradient" that will allow them to evolve continuously and gradually through all of these stages. The approach is that these 10 criteria make up a model. The objective is to evolve toward this ideal by having the largest possible percentage of data that conforms to as many of the characteristics listed above as possible.

Moreover, ethics begins with a collective awareness that involves raising the awareness of each employee and explaining the good practices associated with the digital world. This is why the company as a whole must tend toward being ethic-centric and organize itself around the digital world with a team composed of at least one business representative (data engineer) for the collection, storage and processing of data; a data scientist; a data analyst for the algorithmic part (statistics, machine learning and AI); and a product owner such as a Chief Data Officer (CDO). A data steward can also be present for the strategic aspects associated with ICT.

The entire team will be led and supervised by a Chief Ethics Officer (or Chief Digital Ethics [CDE]) who will provide an overall ethical vision of the desired outcome and act as a link between the stakeholders in terms of ethics.

> DATA SCIENTIST.–
>
> A person whose job involves analyzing data using complex statistical and data mining tools. They produce the value of the data. The data scientist starts with reliable data (thanks to the work of the data steward and has the tools to do it).

> DATA ENGINEER.–
>
> Data engineers are the data professionals who prepare the Big Data infrastructure for processing execution, including those designed by data scientists. They are software development engineers who prepare and commission the hardware and software base that allows for the execution of various optimized data processing operations.

> DATA STEWARD.–
>
> A person who is the administrator or manager of the data. Data stewards have knowledge of the data and are responsible for implementing the strategy in the field. They will apply the governance ordered by the CDO, and ensure that it is followed; the same applies to good practices

and lifecycles. They provide methodological expertise and can ensure the overall consistency of the documentation for a set of data, in coordination with the architects and project owners of the information system.

DATA ANALYST (BUSINESS ANALYST).–

A person who studies data for his or her own business needs, and builds the value of Big Data. Data analysts receive some of the data scientist's work and bring it together with other reporting and other data they have in their possession to do the job. They work with dashboarding, information visualization and information processing tools that are fairly similar to those they use for BI, but they apply them in a different way depending on the data that is made available to them and the business challenges they have to deal with.

DATA OWNER.–

A person who has financial and managerial responsibility for a data set and legally owns it, even if the data set was collected by another party.

Finally, we summarize the different stages of an approach centered on Ethics by Evolution of a digital AI project in Table 3.1.

The structuring of this approach can be articulated by the following series of questions:

– What are the general ethical principles?

– What are the ethics of algorithms?

– What are the reference frame and ethical criteria?

– What are the commitments and articles that make up the ethical charter?

– What are the recommendations and actions that implement the ethical charter?

Ethical principle	Algorithmic ethics	Ethical criteria	Commitment to the ethical charter	Ethical recommendation
Beneficence Autonomy Non-maleficence Justice	Data ethics	6	6	10
	Ethics of systems	7	6	10
	Ethics of algorithms	8	6	10
	Ethics of practice	7	6	10
	Ethics of decisions	8	6	10
4	**5**	**36**	**30**	**50**

Table 3.1. *Steps for integrating an approach to Ethics by Evolution within a digital AI project*

3.3. Temporality relative to the human guarantee in digital technology

We relied on the five stages of ethics that we have seen above, namely ethics of data, ethics of systems, ethics of algorithms, ethics of practice and ethics of decisions.

We have attempted a pragmatic division according to the temporality (before, during and after the design of a device integrating AI), and the actors involved in this human guarantee. For example, in medicine, some AI applications are so complex that explainability is not achievable. Assuming that it is, we could have AI that is completely explainable, and absolutely not ethical! In this case, it becomes essential to have as a recourse human supervision of the machine in the real world by a community of professionals responsible for evaluating the ethical functioning of the algorithmic system: not at every stage, so as not to block innovation, but at critical points throughout the lifecycle, from design to use, after market release. This positive regulation (in other words, a form of "ethical utilitarianism") must be proportionate to the sophistication of the algorithm as well as to its potential dangerousness and the fields of its application.

In order to make this part as pragmatic as possible, we have applied this temporality in the management of care for which AI has an impact on the doctor-patient relationship. This example comes from an explanatory note produced by the Dr. Madeleine Cavet and Jérôme Béranger as part of the working group (GT3) "IA en santé : *Ethics by Design*", co-directed by Brigitte Seroussi and David Grouson, for the digital component of the "Ma santé 2022" plan of the French Ministry of Health.

HUMAN GUARANTEE OF AI.–

The principle of human guarantee of the artificial intelligence device in health must be adhered to. This guarantee must be ensured by, on the one hand, regular verification procedures – targeted and random – of the treatment options offered by the artificial intelligence device and, on the other hand, the development of a capacity to exercise a second human medical look at the request of a patient or a health professional. This second glance can, if necessary, be implemented through telemedicine devices.

In this case, the protagonists concerned by this human guarantee are respectfully the health users, the health personnel and their respective representative bodies.

Thus, we can divide this temporality associated with the human guarantee in digital health according to three major times corresponding to the life stages of the device using AI:

– prior to the design of the device using AI, these are the problems related to data collection;

– during the design of the device using AI, it is the problems brought by the designers, on the device and the algorithms themselves, and their circuit of production and initial validation;

– after the design of the device using AI (use of the device), the issues of use, and impact in real conditions on health personnel and health users who use them, as well as on society. It could also be about the problems of continuous learning of algorithms.

	At the individual level		Representation to designers and decision-makers	
	Health user	**Healthcare personnel**	**Healthcare users**	**Healthcare workers**
Prior to the design of the device using AI				
Health data collection				
Data ethics	Information	Information	GDPR	
Ethics of systems	Security/Rights	Automation	Rights	
Ethics of algorithms			Risk/bias	Protection
Ethics of practice		Training	Justice	Justice/non-discrimination
Ethics of decisions			Participation	Responsibility
During the design of the device using AI				
Tool construction				
Data ethics			Accessibility	Integrity
Ethics of systems			Ergonomics	Privacy settings
Ethics of algorithms			Transparency/explainability	Involvement
Ethics of practices			Privacy by Design	Scientific validation
Ethics of decisions			Autonomy	Human control/explainability
Post-design of the device using AI				
Terms and conditions of use				
Data ethics		Organization	Security	Security
Ethics of systems		Ergonomics	Beneficence	Ergonomics
Ethics of algorithms			Empowerment	Verification
Ethics of practices		Non-maleficence	Transparency/explainability	Transparency/explainability
Ethics of decisions	Empowerment	Free will/benefit	Representation	Scientific validation

Table 3.2. *Temporality associated with the human guarantee in the case of AI in healthcare*

This temporality associated with a digital application illustrates our neo-Darwinian approach of Ethics by Evolution in the continuity of Ethics by Design, which is regularly recommended. This form of evolutionary ethics appears particularly interesting to us in order to be able to evaluate AI on its "moral personality" or "ethical integrity". It consists of considering past experience and integrating evolutionary indicators resulting from a constantly changing environment and sociotechnological context. Henceforth, the criteria of the algorithm and the data corpuses are now one and the same. Such a process further confirms our neo-Darwinian vision, which emphasizes the fact that we make algorithms evolve and grow more than we develop them. Ultimately, the objective is to make an algorithmic system responsible and ethical by integrating criteria of good practice so that it can evolve and enhance its own lines of code throughout its lifecycle.

In our example, a notable complexity in the health field lies in the direct use by users, without the intervention of health personnel, of certain applications or devices using AI on health data or collecting health data[9].

It is, therefore, fundamental that in each life stage and ethical stratum related to AI, all stakeholders (directly or indirectly concerned by the intelligent artificial agent) are involved in order to integrate their requirements. It is only through multidisciplinary consultation that it is possible to obtain an AI that can be morally and ethically acceptable and meet the expectations of health professionals or their representations, on the one hand, and health users or their representations, on the other hand (see Table 3.2).

3.4. For the health user and for health user representation

Prior to the design of the device using AI:

– data ethics:

- individual user: information on the collection of health data and its purposes, free and informed consent (information),

- user representation: GDPR compliance;

9 In areas as varied as well-being, sports, monitoring of chronic diseases, community applications and social networks, mutual insurance, etc.

– ethics of systems:

- individual user: information and reality of data security, right of withdrawal of consent, right of opposition, etc. (security/rights),

- representation of users: rights of withdrawal of consent and opposition (rights);

– ethics of algorithms: user representation: reduction of unnecessary or miscalculated risks as well as unfair biases and discriminations (risk/bias);

– ethics of practices: representation of users: equality and/or fairness of treatment and guarantee of respect for the fundamental rights and freedoms of all citizens (justice);

– ethics of decisions: user representation: integration of the health user in the design meetings of the digital project associated with AI (for the representation of health users) (participation).

During the design of the device using AI:

– data ethics: user representation: same rule of access and diffusion of information, regardless of the health user's profile or status (accessibility);

– ethics of systems: user representation: ergonomic simulation interfaces so that they can appropriate the functioning and effects of the algorithm (ergonomics);

– ethics of algorithms: user representation: transparency and explainability to the patient on the description of algorithmic operations involving the AI source code (transparency/explainability);

– ethics of practices: user representation: respect and protection of the privacy of individuals by integrating a security policy relating to patient confidentiality (Privacy by Design);

– ethics of decisions: user representation: free and informed consent and adherence of the individuals or institutions at the origin of the digital data that feed AI (autonomy).

Post-design of the device using AI:

– data ethics: user representation: duty of security and protection of digital data (security);

– ethics of systems: representation of users: appropriateness and social benefit for the individual, or even the community and humanity (beneficence);

– ethics of algorithms: user representation: verification that the increasing and evolving autonomization of AI always serves the interest of the individual, or even the general interest of society (empowerment);

– ethics of practices:

- individual user: explainability and transparency, fair and adapted information (transparency/explainability),

- user representation: explainability and transparency (transparency/ explainability);

– ethics of decisions:

- individual user: individual benefit for the user, users at the center of their decision (empowerment),

- representation of users: criticism and putting into perspective of results, citizen participation in social debates on AI (representation).

> EMPOWERMENT.–
>
> The user could become more proactive by asking platform operators to make their algorithms more transparent, having control over their personal data, migrating to platforms that have a strong commitment to transparency in their operations, or supporting and funding new digital projects based on principles of transparency and ethics.

3.5. For health personnel and for the representation of health personnel

Prior to the design of the device using AI:

– data ethics: individual health personnel: readable information summarized for the caregivers on the legal framework of data management, good practices, purposes of processing (information);

– ethics of systems: individual health personnel: technical and scientific relevance of the tool automation;

– ethics of algorithms: healthcare staff representation: data protection policy during the operational life of AI (protection);

– ethics of practice:

- individual health personnel: initial and continuing education of caregivers on AI issues (training),

- health worker representation: equal treatment and non-discrimination (justice/non-discrimination);

– ethics of decisions: representation of healthcare personnel: setting up an inventory of the professional objectives of professionals throughout the lifecycle of the data (input, during, output) in order to list and indirectly trace responsibilities (accountability).

During the design of the device using AI:

– data ethics: representation of healthcare personnel: ensuring the quality, integrity and choice of information transmitted to the individual in accordance with the best standards (integrity);

– ethics of systems: representation of health personnel: confidentiality parameters and measures around the architecture and structure of AI (confidentiality parameterization);

– ethics of algorithms: representation of healthcare personnel: involvement of caregivers in the design and validation of algorithms (involvement);

– ethics of practices: representation of healthcare personnel: scientific validation of *in vitro* algorithms, transversality of the people involved in the digital AI project (scientific validation);

– ethics of decisions: representation of health personnel: accompaniment of the decision established by AI of a satisfactory explanation, or even control by the human being, provision of a clear explanatory framework for the decision-making of the health professional) (human control/ explainability).

Post-design of the device using AI:

– data ethics:

– individual health personnel: relevance of the tool: matching the use of AI and therefore the use of digital data with the organization of the structure that uses AI (organization);

– representation of health personnel: safety and protection of digital health data (security);

– ethics of systems:

- individual health personnel: ergonomics, non-multiplication of work tools (ergonomics);

- representation of healthcare personnel: ergonomics, interoperability (ergonomics);

– ethics of algorithms: health workforce representation: verification that the increasing and evolving empowerment of AI always serves the interest of the health user (verification);

– ethics of practice:

- individual health personnel: non-maleficence, support to help the health personnel in explaining to the patient (transparency/explainability);

- representation of health personnel: explainability and transparency (transparency/explainability);

– ethics of decisions:

- individual health personnel: right to derogate from the suggestion of the device integrating AI, individual benefit for the user (for the health personnel) (free will/benefit);

- representation of health personnel: clinical validation of AI devices in real conditions (scientific validation).

In addition, the implementation of a management supporting AI implies acting on four specific and complementary areas:

– the evolution of the type of piloting, by setting up a process of change with the aim of mastering all the systems;

– improvement and/or development of the methodology, according to the different levels of management and use of AI for strategy, operations, or performance;

– the evolution of IT resources with the reliability, robustness, and completeness of the information, and on the other hand the adjustment of AI to make it possible to develop BI computing to complement production computing;

– changes in the organization, by bringing greater transversality and control of the human resources employed by adjusting the skills and the organizational structure.

Such an approach to managing technological innovations can only be achieved through rigorous project management, where these four areas of action must be carried out jointly and simultaneously. From now on, companies must develop and structure a real management strategy around the integration of an AI. It is becoming necessary for them to be able to clarify their priorities and, therefore, identify precisely the actions that will need to be developed, produced, or reduced, or even eliminated. The management is only the support of the strategy in action.

Indeed, project management must lead to a common and unifying dynamic, a real multidisciplinary exchange. In this leadership context, digital devices are the vector through which information circulates and is disseminated among all levels of the organization. Generally speaking, it can be seen that the inter-connections between the different sectors of activity are often subject to numerous dysfunctions. Reducing the number of these dysfunctions generally leads to improving the fluidity of work processes, transmission, access to information and consumption of resources, while respecting social and moral obligations such as confidentiality and autonomy.

Thus, AI must also be studied in the context of each structure that disposes of its own history, its own strategy and its own axes of progress. The aim is to be able to follow the direction and objectives that the structure has defined for itself. The whole point of ethical management is to bring meaning and purpose to the organization that has been set up (see Table 3.3).

Moreover, the digital environment is by nature polymorphic, heterogeneous and multidisciplinary. Consequently, the person in charge of conducting digital governance within a structure must possess these characteristics. This is a necessity if one wishes to carry out interdisciplinary steering on the principles and standards that control the use of digital technology. Refocusing on the ethics of AI has the virtue of saving them from themselves and their excesses in order to move toward greater transparency, trust, sincerity and acceptability. To do this, companies are led to create and integrate a new function within their organization chart, the Chief Digital Ethics (CDE).

Types of measures	Questions
Vision	What does the organization want to become?
Mission	What is the reason for the structure's existence?
Values	What are the principles of conduct and decision of the company?
Medium-term strategic axes	What are the medium-term strategic axes and their coherence?
Annual strategic highlights	What are the strategic priorities that need to be present to everyone?
Overall objectives	Are the company's objectives in line with the digital environment and the market?
Objectives by process	What are the annual performance objectives and progress in digital processes?
Objectives per team	What are the collective objectives and targets being managed at the first level of leadership?
Individual goals	What are the singular objectives or contributions of each person to the objectives of his or her team?

Table 3.3. *Strategic launching of AI within a structure*

Finally, based on this observation, four success factors for the implementation of an AI strategy can be presented:

– a strong involvement of the company's general management in order to allow optimal adherence and visibility on the part of all employees. This helps to reinforce an "AI culture";

– an indispensable cross-disciplinary approach, from the very beginning of the implementation of an AI project. To do this, it is essential to mobilize departments that have different but complementary points of view. The notion of multi-disciplinarity is, therefore, crucial along the entire value chain resulting from data exploitation. It is essential to try to communicate as much as possible internally;

– trust by design/evolution, based on a consideration, upstream and throughout the digital project, of the ethical and legal issues associated with AI until its implementation and use. The guarantee of security, confidentiality, integrity and use of processed data is, therefore, a prerequisite to enhance the value of personal data. The structure must ask itself a series of questions in order to make the use of AI operational and efficient (see Table 3.4);

– the implementation of an analysis program, in which it is important to devote time to identifying the company's needs and challenges and to its maturity diagnosis, and only then to select the technologies best able to meet the identified and qualified needs. From then on, AI responds to the major challenges for the company, which are to increase the capacity to support its own activity, to gain in productivity, but above all to innovate compared to the competition.

TEST AND LEARN.–

A set of practices followed by retailers, banks and other consumer-focused businesses to test ideas in a small number of sites or customers to predict impact.

ROI.–

Return on investment (ROI) is a financial ratio that measures the amount of money earned or lost in relation to the amount initially invested in an investment.

Questions	Actions to be carried out
What information do we need to innovate and be competitive?	Identify the business opportunities offered by large volumes of data
	Draw inspiration from innovative initiatives in the sector
What is the nature of AI?	List the primary function of the data (categorize, calculate, collect, measure, etc.).
	Describe the epistemological nature through their support and mode of supply (numbers, codes, tables, texts, databases, etc.).

Questions	Actions to be carried out
How can data that contains personal information be secured within AI?	Put in place means to ensure that the transformation is compatible with free and informed consent and anonymization of the owner of the data
	Ensure the technical relevance and human merits of the tool, reducing unnecessary or miscalculated risks
How can Big Data be integrated with existing data flows and repositories that feed AI?	Identify and list the common areas between Big Data and existing data used by AI
	Establish a single purpose for AI in which Big Data complements and articulates existing data
What are the undertapped and untapped data at our disposal?	Know the available data sources of the company
	Know how to interpret raw data
Which legal framework regulates the exploration of Big Data by AI?	Make an inventory of the regulations relating to the processing of "massive data"
	Establish an internal guide to good practices, deontological code or ethical charter based on this exploitation resulting from AI
What are the major data to be managed by AI?	Describe the intrinsic qualities of the data (applicability, ordered, reliable, relevant, universal, federative, etc.)?
	Search for legitimate and referent sources of the data
Are we ready to "extract" useful information from our data?	Have the skills to analyze data (including data scientists and data analysts)
	Frame the associated organizational, ethical and regulatory transformations
What is the validation system data applied internally?	Build an internal data validation process (validation criteria and metrics)
	Define the profile of the validator, the nature of the tools (if the validation is automatic) and the frequency of the device
How can a traceability of the data used by AI be obtained?	Build a visualization dashboard that makes it possible to follow the whole data lifecycle (mapping)
	List the different transformations performed and identify the intermediate data produced
Who in the organization has access to customer data?	Establish at each step of the process of data processing, a list of people who have access to customer data according to their professional status

<table>
<tr><td>What monitoring systems are in place to track potential violations and damage that the exploitation of data by AI can be subject to?</td><td>Carry out a mapping of the different processes that allows for monitoring potential damages and violations around the processing of AI data</td></tr>
<tr><td rowspan="2">What management and strategy are to be put in place to carry out a digital project related to AI?</td><td>Establish a clear and flexible action plan</td></tr>
<tr><td>Promote the “board” mode by bringing together people in the same physical space</td></tr>
<tr><td rowspan="2">Which management can be put in place to successfully complete a digital project?</td><td>Develop and implement new collaborative strategies[10] centered on AI</td></tr>
<tr><td>Working in startup mode, i.e. with great agility and speed of execution, simplified prototyping, test and learn, and also the right to make mistakes</td></tr>
<tr><td>Who in your organization has access to customer data?</td><td>Do not always lean on and attach to an ROI</td></tr>
<tr><td>What monitoring systems are in place to track potential violations and damage that the exploitation of the data may be subject to?</td><td>Stimulate and obtain the acceptance of users[11] for the implementation and realization of the digital project around AI</td></tr>
<tr><td>What monitoring systems are in place to track potential violations and damage that the exploitation of the data may be subject to?</td><td>To set up the conditions for a dynamic of co-innovation[12] around AI</td></tr>
<tr><td>What monitoring systems are in place to track potential violations and damage that the exploitation of the data may be subject to?</td><td>Collaborate with startups to move forward faster</td></tr>
</table>

Table 3.4. *Questions and actions to prepare an AI-related project*

3.6. Environmental parameters of digital technology

Today, digital technology is omnipresent in our society. All sectors of activity are impacted by this “digital tsunami”, which continues to grow with

10 The development of ecosystem and platform logics, the value chain and the modes of capturing the value of a platform, the conditions for controlling the associated risks, the case study of ecosystems, etc., will be discussed.

11 The willingness of citizens to transmit their personal data, the conditions and actions to establish a relationship of trust and support for the role.

12 The data governance strategy, internal organizational and cultural projects, co-designing solutions with users (main determinants, contributors’ motivation axes), etc., are all part of the process.

time and technological innovations. A decision that is generated by or with the help of technology can be justified on a technical level, but be declared inadequate in terms of a legal aspect or impossible to implement on a logistical level (reality principle).

In doing so, it must be considered that the ecosystem and the situation are dynamic, and temporal considerations are fundamental. Thus, events, some of which cannot be predicted, are linked together over time and constitute a complex system within which it is essential to make decisions in real time, considering the evolution of the environment and the sources of knowledge and information available.

Moreover, if we look at the history of NICTs, we can see that it can extend over four very specific successive eras (Dherse and Minguet 2007; Ponçon 2009; Béranger 2015):

– the technical era, representing mass production;

– the organizational and regulatory era, which tends toward an optimal contribution of NICTs to the performance of business processes;

– the behavioral era, reflecting the relationship and trust of professionals toward society;

– the era of purpose and ethics, expressing the accountability of the actors directly or indirectly involved in these NICTs.

On the other hand, we can study the digital environment – especially AI – from a prism and a sociocognitive framework that can be declined in three dimensions (Vayre 2017):

– the learning environment that designates all the devices for collecting and structuring data that allow the machine to build a representation of the world (databases);

– the processing environment that designates all the devices that allow the machine to make inferences about the world from the learning data at its disposal ("black boxes" of deep neural networks);

– the political environment that designates all the evaluation devices that enable the machine to self-regulate its learning (regulation and digital governance).

DEEP NEURAL NETWORKS.–

These multi-layered neural networks can contain millions of neurons, divided into several dozen layers. They are used in deep learning to design supervised and unsupervised learning mechanisms.

Moreover, beyond the fact that digital technology integrates all areas of activity in our daily lives, it also has an impact in all the sectors that make up a company. Based on this observation, we can divide the environmental parameters of NICTs, particularly within a company, into eight categories (see Box 3.2).

Structural: The way in which the internal parts of a whole are arranged together. This intrinsic value of a system gives it its coherence and is its permanent characteristic.

Technological: This corresponds to the ecosystem or data sphere in which NICTs gravitate. It includes the network of partners (hosts, publishers, datacenters, operators, Pure Players, Cloud Computing, etc.), the various key activities (consulting, computing, analysis and storage) and offerings (collection, processing, visualization, interpretation, storage, etc.), resources and distribution channels (business to business [B2B] and administration to business [A2B]), cost structures (maintenance, subscription, operation, development) and revenue streams (valuation, sales, transacting, licensing, intellectual property, rentals).

Strategic: It is the set of coordinated actions, operations, maneuvers and the procedure to follow in order to reach a specific objective. Strategy is largely influenced by the economic context and by the people who build it.

Methodological: This represents the logically ordered set of steps, principles and actions that constitute a means to an end. It is, therefore, the way of carrying out, according to a reasoned and coherent approach, a work or an activity.

Organizational: This refers to the action of structuring, delimiting, articulating, arranging or distributing. It can be seen as a social process.

Regulatory: Everything concerning regulations, laws, standards and measures, certification and labeling schemes. This sector imposes a particular rigor in its application within the organization.

Relational: This is the set of interactions, exchanges and sharing that exists between the various actors directly or indirectly involved in NICTs.

Cultural: This constitutes the behaviors, habits and morals, uses, practices, perceptions and approaches of users of technological tools.

Box 3.2. *Environmental parameters of NICTs within a company*

Thus, the digital environment or ecosystem is based on these eight domains that articulate and interact with each other. These parameters should not be isolated and static from each other in order to consider the complexity of reality. Indeed, a good analysis of the environment should consider these domains as constituent parts of a single whole. These environmental parameters will be the structural basis of our reflections on AI regulation and governance.

Let us note, finally, that in the perspective of a progressive emergence of applications associated with quantum mechanics (in particular quantum computing and quantum teleportation) marking the beginning of the fifth industrial revolution, it will be fundamental to integrate temporal metrics into these eight parameters in order to complete the digital ecosystem around NICTs.

3.7. Regulation associated with AI

Within the framework of NICTs, ethics targets acts and facts that have an incomparable causal impact on the future and that are accompanied by a predictive knowledge that, however incomplete it may be, goes beyond anything we have known before. To this must be added the order of magnitude of long-term actions and very often also their irreversibility. Responsibility is then positioned at the heart of ethics, including the spatiotemporal parameters associated with the four ethical principles, as seen above. Thus, the algorithmic responsibility of AI can be established by ensuring that the launch of machines is ethically aligned with these universal principles and associated moral values. Under these conditions, it is necessary to identify which ethical principles an intelligent machine must follow, and according to which situations. This leads us to believe that a close relationship between governments and companies will have to be defined. We must never lose sight of the fact that currently the so-called "weak" AI only does what its designer tells it to do. It is then up to governments, software publishers and companies to make an explicit and precise choice when developing the artificial agent and when implementing responses to different situations. How can all this be harmonized, knowing that, depending on the history, culture, politics and jurisdiction of a State, because each country has its own technology and its own ethics, and different parameters for making autonomous decisions? How will these ethical decisions be made at the level of the entire planet? Will they have the same moral specifications? Finally, it comes down to this question: How can future AI be made more humane and inclusive?

In this context, the regulatory perimeter must take these questions into consideration to determine where the algorithmic responsibility for the decision-making carried out by AI begins and ends. In my opinion, it would be preferable for an international regulatory authority to decide on the ethical principles and rules to be considered. This body will have to be the guarantor of the correct application of this base of ethical measures in technologies by a process of evaluation, audit and rating. It is clear that machines need a global agreement on ethics in AI for the purpose of standardizing the means of algorithmic accountability and responding to ethical dilemmas.

Therefore, the pace of innovation in AI requires a constant effort of self-regulation, regulation, dissemination of best practices, and control by trusted third-party organizations, for example, in collaboration with all stakeholders, including levels of government or professional bodies. Regulation must be one of the keys to the successful adoption of AI. Systems need to be cashed in to gain acceptance and be able to grow. This regulation around algorithmic processing incorporates five challenges per environmental parameter related to cloud computing (see sections 3.7.1 to 3.7.8).

Finally, with regard to the regulations governing AI, we note that this legislative framework could be improved for many reasons:

– first, it is essential to ensure that the existing regulatory framework considers a set of ethical principles that are essential to building safe and caring AI;

– second, at present, there is a risk of confrontation and division due to national divergences in the application of existing rules. Hence, the idea of setting up control bodies (regulators, or even the European legislator) in order to progressively guarantee a uniform interpretation of the existing rules;

– third, depending on certain sectors, existing regulations may not be sufficient. It would then be possible to adapt the existing legislative framework, or even adopt new rules, in order to respond to a number of risks and specific situations. For example, to improve the legal liability regime with a view to ensuring a more effective and fairer system of compensation for damages caused by the use of AI.

3.7.1. *Structural parameter*

The structural parameters are as follows:

1) the structuring of merged data from distinct sources for a more exhaustive analysis;

2) the compliance of the algorithmic devices concerned;

3) the control of interconnections and interactions between the different processing systems, strategies and policies for the enhancement of digital data;

4) continuous improvement that is applied to data quality;

5) the development of reliable results that reflect the reality at time *t* and for a given situation.

3.7.2. *Technological parameter*

The technological parameters are as follows:

1) preservation and recycling of digital data in the environment;

2) the establishment of transparency and readability through better explanation and intelligibility of the rules and assumptions underlying the functionalities, the data that come in and go out and the objectives of the algorithmic processing. This operating logic must be stated verbally – and not as lines of code – in order to be understood by all human beings;

3) consideration of the design of algorithmic systems; in other words, the human-machine interface, in order to counterbalance the "black box" effect

and informational asymmetry of algorithms, by reinforcing the autonomy and reflexivity of individuals (data visualization);

4) the maintenance of a permanent effort of intelligibility to restore confidence by explaining (pedagogy and awareness) – making oneself understood (giving access to parameters, goals, causes, issues) – notifying oneself (informing that the person has been measured);

5) the anonymization of numerical data from the source code.

3.7.3. *Strategic parameter*

The strategic parameters are as follows:

1) the assurance that forms of human deliberation surround and support the use of the algorithms, through supervision and audit of the parameterizations, and also of all the effects – direct and indirect – of the latter;

2) the possibility of reforming the principles governing the use of algorithms, following national consultation, in order to respond to changes in society (mutability);

3) financial encouragement, at the national and international level (like the European Union (EU)) for the development and use of AI technologies that are socially preferable (not just acceptable) and environmentally friendly (not only sustainable, but also eco-friendly), sustained, increased and coherent European research, an effort around the ethical approach to AI, research on public perception and understanding of AI and its applications, and the establishment of structured public consultations mechanisms for designing policies and rules related to AI;

4) the reduction of categorization and discrimination in society;

5) the policy of openness of numerical data that is implemented internally and externally of a structure (open data/publication of data and/or algorithms).

3.7.4. *Methodological parameter*

The methodological parameters are as follows:

1) the management of accessibility to digital data;

2) supporting the development of self-regulatory codes of conduct for data and AI, with specific ethical duties according to the professions. This would be in line with other social groups of sensitive professions, such as doctors or lawyers, i.e. with the supporting certification of ethical AI through trust labels to ensure that individuals understand the interest of responsible AI and will, therefore, demand it from providers;

3) the integration of ethics in the practices carried out throughout the chain of development of digital data;

4) the development of a repair process or mechanism[13] to repair or compensate for an error or fault caused by AI. To foster public confidence in AI, society needs a widely accessible system and a reliable redress mechanism for the harm, costs and other grievances caused by the technology;

5) the implementation of ethical rules on the use and exploitation of data that are defined, applied and shared internally, and communicated to the persons concerned (end customers and/or partners).

3.7.5. *Organizational parameter*

The organizational parameters are as follows:

1) management and traceability of the lifecycle of an analytical tool;

2) clearly defined organization and roles for digital data management;

3) the (prospective) management of the required skills around the data is in place;

4) the systematic mediation of an independent trusted third party during any conflict concerning the launch of an AI tool or algorithm;

5) the development of criminal liability and specific sanctions. Within the framework of the *ex ante* conformity control, the civil liability of the company may be engaged – or even the criminal liability of the project director, its technical director, or its Chief Ethics Officer (or CDE) – if it is proved that the existence of significant differences between the algorithmic

13 Such a mechanism will necessarily imply, on the one hand, a clear and precise strategy and, on the other hand, a full attribution of responsibility to humans and/or organizations. AI designed to be infinitely self-developing or self-replicating, at the risk of becoming very numerous or very advanced rapidly, must be subject to rigorous security control.

processing and its data sheet is due to intentional fault. Otherwise, the company will be exposed to reputational risk. Within the framework of an *ex post* control of lawfulness, if the processing is judged to be unfair, it will then be the responsibility of the audit structure, which will be questioned according to the same modality, or that of the person in charge of the public service mission, in the event that the algorithmic processing is compliant and fair, but not its use.

3.7.6. *Regulatory parameter*

The regulatory parameters are as follows:

1) encouraging regular reviews of legislation to determine the extent to which it promotes socially positive innovation in AI. The latter must be subject to the same legal rules that apply to human beings. Opposition to the development and use of AI that would violate international conventions or human rights and the promotion of safeguards and technologies that would not harm;

2) the supply of personal data of European nationals to AI must comply with the requirements of the GDPR[14] (Privacy by Design). Therefore, each individual should have the right to access, manage and control their personal data, because of the power of AI systems to analyze and use this data. Any citizen who believes that his or her rights and freedoms are not fully respected, or would likely not be respected through the use of algorithms, as well as any person witnessing the use of an algorithm that appears to be inconsistent with the principles set forth, should have the right to the lawfulness of the algorithm and its use evaluated *ex post* by a commission of experts (right of appeal);

3) the establishment of soft law and "platform state" mechanisms. The rules relating to the control of algorithmic processing programs should be hard law. Ethical principles should be soft law and co-created by government departments and civil society. An independent administrative authority will act as a link between the public authorities and program designers in order to encourage technical innovation. Algorithmic processing

14 Regulation 2016/679 is a European Union regulation that constitutes the reference text on the protection of personal data. It strengthens and unifies data protection for European Union nationals.

must offer open APIs (Application Programming Interface) to allow any citizen to develop models or new programs based on these treatments;

4) the development of appropriate legal procedures and the improvement of the computer infrastructure of the justice system to enable the scrutiny of algorithmic decisions in court. This would likely include the creation of an AI explainability framework;

5) the opening of summary proceedings before the Defender of Rights and Freedoms of France, for example. The Defender of Rights and Freedoms will be the competent authority to receive citizens' petitions in order to bring them before the aforementioned auditing authority. It will set up a dedicated and simplified referral channel for this purpose.

3.7.7. *Relational parameter*

The relational parameters are as follows:

1) the responsibility of the actors involved directly or indirectly by the NICTs;

2) the free and informed consent of the person concerned by the personal data;

3) the development of a constructive exchange between AI developers and legislators;

4) the respect of the expectations of the citizens and the requirements of the digital professionals;

5) the guarantee of reciprocity in the sharing and accessibility of data to citizens in order to reduce societal inequalities.

3.7.8. *Cultural parameter*

The cultural parameters are as follows:

1) understanding and adapting advanced analytical environments;

2) education, awareness and listening to human beings, and active engagement of stakeholders to seek their feedback on attentions, to inform them and to answer their questions about AI. Consequently, the overall functioning of AI must be understandable and interpretable by people in

order to be able to explain it. Every person must have the right to be educated and to be able to develop mentally, emotionally and economically as a complement to AI. This can be achieved by supporting the creation of school programs and public awareness activities around the societal, legal and ethical impact of AI;

3) supporting the ability of corporate boards of directors to take responsibility for the ethical implications of corporate AI technologies. For example, this may include improving the training of existing boards and the potential development of an ethics committee with internal audit powers;

4) the assurance of continuous training and information of the actors. Training seminars will be set up to raise awareness among the technical staff of private and public structures, developing or using algorithmic processing, on the ethical issues they raise and the legal framework to which they are subject. These will take place within the professional structures, and not upstream, in the training cycles of engineers;

5) cultural and socioprofessional diversification through the involvement of people in the elaboration of algorithmic systems in order to avoid some form of ethnocentrism.

> DATA VISUALIZATION.–
>
> Data visualization is one of the basic criteria in the success of Big Data processing. This dataviz (or data visualization) has developed at the crossroads of design and statistics. It constitutes a structuring and collaborative approach to support the data produced by the connected objects. Its added value lies in the representation or personalization of the data and the dissemination of its content to operational decision-makers and the general public so that Big Data is considered useful by the latter. This data visualization is, therefore, both an analysis and a readable graphical formatting, in particular via dashboards or radar representations.

Moreover, AI and Big Data are of no use if no one can take advantage of them or if they are not exploited via an automated device designed by individuals. The integration of digital data aims at simplifying as much as possible the access to, understanding, and processing of data. Faster turnaround times and the elimination of bottlenecks due to lack of skills and smooth interactions allow companies to become faster and more efficient. From now on, the new sources of value creation around the exploitation of

Big Data by AI will be refocused on a limited number of data management platforms. Based on this, three levels of strategic challenges are revealed:

– proceed to a controlled opening of AI architectures;

– establish a data governance that builds trust internally and externally;

– bet on triggering AI network effects.

Consequently, some of the ethical issues associated with ICT are conducted with reference to the policy, strategy and methodology employed by the company. Infrastructures should be required to adopt appropriate policy and code management mechanisms to avoid ethical issues that may arise. Prior to any strategic development and decision, it is common practice for a company to establish a SWOT[15] table to obtain a study of the relevance, the stakes and the coherence of future action. Based on this observation, we have developed a SWOT around the issues of AI for a company. Our SWOT matrix is composed of 10 items illustrating strengths, weaknesses, opportunities and threats, respectively (see Table 3.5).

Strengths	Weaknesses
Development of analytical tools efficient in the service of the structure	Slow organizational change management
Emergence of new services for the company	Lack of skills to manage AI
Reinforcement of the competitiveness between industrialists	Opacity on the operation of AI
Predictive analysis via processing algorithms	Reliability, integrity and construction bias data used by AI
Development of analytical environments	Control and traceability of the massive data that feeds AI
Creation of a unique and reusable version of reality	Obtaining free and informed consent of the data owner
Understanding and anticipating behaviors and needs and customer expectations	Limited data retention
Personalization of offers (finer segmentation and targeting)	Data access control
Participatory, preventive and predictive services (Industry 4.0)	Awareness and training in an "AI culture"
Automation of decision-making processes	Changing work paradigm

15 SWOT analysis is a corporate strategy tool whose mission is to determine the strategic options available at the level of a strategic business area. SWOT is an acronym for strengths, weaknesses, opportunities and threats.

Opportunities	Threats
Generalization of AI in all the spheres of society	Unclear regulatory framework and limited regulation
Calls for public research projects around AI	Unethical internal and external uses of AI
Reduction of expenses and costs	Non-adherence to the rights of the citizen (right to information, oblivion, opposition, access, and rectification)
Optimization of the customer's journey	Disempowerment of professionals
Fast detection of weak signals at a given event	Security and protection of information
Development of self-monitoring (management tool)	Discrimination, categorization and individualization of society
Realization of decision support solutions (expert system)	Non-adherence to privacy and individual liberties
Better understanding of mechanisms, processes and even events	New wave of socioeconomic inequalities and injustices
Continuous process improvement	Practice development and a less humane decision

Table 3.5. *SWOT table of AI issues for a company*

Therefore, companies must make human and material investments related to a strategy and a policy established around AI. Those who will be able to take advantage of their intangible capital will open new perspectives toward more competitiveness and innovation. This is why infrastructures are considering new methods of analysis and management of unexplored data in order to bring them added value. A series of questions arise: is this a real opportunity and what are the challenges of AI for them? What choice should be made? Should we accumulate data processing systems or take a device that includes all the features that meet present and future needs? Should we favor technicality or ergonomics and flexibility? How do we obtain a coherent whole?

In the light of this questioning, it appears essential that the strategy and methodology for integrating Big Data used by AI must consider the existing forms and sources of data in a new device that supports all phases of a data supply chain. Big Data can only be valuable if it is integrated and merged with existing data. However, this does not mean abandoning current data integration and business intelligence (BI) methods, as they must be integrated with newer solutions that support the use of Big Data. The optimization of the ROI of companies is done through a highly customizable analytical solution that allows for the merging of Big Data with their BI and application infrastructure. As a result, data integration is driven by the need for analytics to enable growth and adaptation. Integration and the study of these gigantic data volumes go hand in hand and are generally inseparable. This digital turn around AI must be done for companies in a progressive way after considering the context, impacts and changes in the environmental parameters that this entails. Analytical, visualization and BI processes need to be in phase and interlinked with each other in order to move toward a common goal. Poor integration irremediably leads to a limited value of AI.

In addition, if we take the AI report[16], developed in 2017 by Accenture Research in collaboration with Frontier Economics, we can identify eight good practices for implementing a people-centered approach to AI in companies:

– defining an AI-oriented strategy: to benefit from the profitability generated by AI, decision-makers within companies must recognize its effectiveness and act in order to realize the benefits;

– reinventing human resources by integrating AI: the role of the human resources manager will not only be to look after employees, but also to manage the human-machine interactions associated with AI;

– learning with machines: in order to adapt their activities to the changing nature of employee learning and training, managers must focus on the needs of their staff, particularly with regard to the development of agile skills;

– appointing a Data Production Chain Leader: this is a necessary position to build a complete integrated supply chain;

– creating an open artificial intelligence culture: trust, openness and transparency are essential to the smooth running of human-machine relations and leaders are invited to design a corporate culture and guidelines to reduce the risks associated with a mixed workforce while optimizing opportunities;

16 Accenture Research. Embracing Artificial Intelligence: Enabling Strong and Inclusive AI-Driven Economic Growth. Updated December 2017.

– optimizing user data in the cloud: the next phase of innovation will combine user-enriched data in the cloud with AI capabilities to create new business opportunities;

– a step beyond automation: with recent advances in AI, companies must take further steps to harness the intelligence of autonomous, dynamic and self-learning machines;

– measuring the return on AI: unlike traditional assets that lose value over time, AI assets gain value and financial managers will need new financial parameters to properly assess the "return on AI", which could include the value generated from each algorithm or a combination of initial investments and current expenditures.

Finally, no one can decide alone what is good or bad. There is an urgent need to discuss the directions we want to take as widely as possible. There is a danger in exchanging AI among an elite and that ordinary people feel left out, or think that we decide things that are not good for them. Researchers have a responsibility to explain the issues to the public, to governments and to businesses. But in developing benchmarks, all actors in society are involved, especially businesses (which develop applications based on the work of researchers) and citizens.

3.8. Algorithmic systems and digital data governance

Algorithms have entered our daily lives without our consent and modulate our relationship to the world without our really knowing about their existence, the extent of their action, their power and the criteria they use to decide our existence for us.

One aspect of the Big Data revolution is based on more powerful and accessible technologies and the mathematical power of algorithms to make data talk. These "megadata" pose a major challenge in terms of adaptation and upgrading that requires specific management and supervision. The challenge is to avoid being overtaken by isolated initiatives carried out by the "business departments" of a company tempted by Big Data, without really paying attention to the 7Vs (volume, variety, velocity, veracity, visualization, volatility and value) of these new forms of data, as well as to their security. The mission of this algorithmic governance is, on the one hand, to optimize the management of information risks, and on the other hand, to improve the structure's usage behavior and activity. It develops leverage effects that promote the reinforcement and fluidity of the systems relating to the development of the activity or services offered by a company or structure. The strength of governance is, therefore, a cross-functional approach based on an overall view of the lifecycle of "massive data" in order to control all the layers of the organization and guarantee the efficiency of processes.

Indeed, several examples reveal how a malicious or accidental use of Big Data technologies can transform an algorithm into a silent and systemic tool to be discriminated against. This is why data and algorithms must be the target of well-thought-out and well-defined governance rules, for example:

– having a management policy around AI that uses data based on a limited and known number of principles. This change management program must be carried out over the long term;

– applying governance methods to AI priority data;

– analyzing the structure of a source and defining the algorithmic exploitation modes of this data according to its purpose (business) or its potential[17] in order to use it properly;

– evaluating the reliability rate and quality of information[18] in order to optimize its use;

– introducing "information and AI protection officers";

– setting up steering committees for the management and development of data within the company;

– writing internal procedures and guides of good practices, on the one hand, on the protection and security of information, and, on the other hand, on the operation of AI;

– not neglecting the human (behavior and education) and organizational dimension;

17 This potential is complex to comrehend, since generally the value is revealed by coupling several types of information together. It is, therefore, essential to evaluate this potential in relation to other available sources.

18 The quality of Big Data sources has a direct impact on the usability of the information and the cumbersome "clean-up" processes associated with it.

– not making decisions on obsolete information;

– implementing a data reduction strategy, where only data with a defined value are kept by AI;

– not retaining information longer than necessary.

In summary, information and AI governance promotes transparency in the management and use of intangible capital by organizations and companies. The challenge is, therefore, to build trust and better value the functional contributions within a structure.

This governance is, therefore, becoming an essential strategy for Enterprise 2.0, which wants to move toward a 4.0 model that must meet the following functionalities:

– restore: search and visualization functions, semantic and spontaneous search, data mining;

– understand: evaluation, creation and extraction of raw data, contextualization and security, management of the development cycle;

– enhance: the selective distribution (or push) of data to applications such as mashups, the exchange of content through the Web and social networks, the collection of information in its environment:

- control: traceability of devices and accesses (confidence), data deduplication, deletion of irrelevant data and content,

- imagine: personalization of services, socialization of the company, adaptation of IS,

- win: speed, compliance, agility and efficiency,

- serve: transparency of use, independent availability of technical support (tablet, computer, smartphone, etc.), content integrity and certification assurance,

- coordinate: the modes of interaction of the actors in the management of healthcare,

- unite: structured and unstructured data, NICT applications, public and private cloud computing, business value, and asset value,

- coach: the decision-making and action capacities (empowerment) of users (consumers, citizens, patients or health professionals),

- anticipate: the interoperability of solutions, the automation of exchanges between connected objects, changes in behavior and habits,

- certify: trust around the use of data.

> DATA MINING.–
>
> In a marketing context, data mining brings together all the technologies likely to analyze the information in a marketing database in order to find useful information for marketing actions and possible meaningful and usable correlations between the data. On a more general level, data mining is a process that enables the extraction of commercially relevant information from a large mass of information.

> PUSH.–
>
> Push refers to the process of delivering information to individuals, without the latter having to search for it by their own means. It is to be linked to the concept of push and pull. Its main aim is to push a product toward the individuals rather than to attract the individuals toward the product.

Moreover, governing also means guiding, orienting, setting the course and following an adapted implementation. The data governance process must, therefore, organize all data related decisions in search of:

– enhancement of heritage;

– maximization of the use of these data;

– quality optimization.

Moreover, information governance impacts all market players such as:

– solution vendors and integrators who are obliged to have their offers evoked in order to respond to market developments;

– companies and people who carry out information flows for the whole organization;

– the consultants who support the digitization of companies.

Information and knowledge management has now become a major business challenge for:

– sharing and enhancing the data and information flows that form the intellectual foundation and part of the intangible capital of the organization;

– increasing sales (products and/or services) and marketing performance;

– promoting and encouraging innovation;

– improving administrative efficiency and reducing costs;

– ensuring regulatory compliance.

Therefore, this observation and analysis establishes a series of measures and approaches around AI that must be:

– accentuated: anticipation of needs and requirements, knowledge of the citizen, personalization of follow-up and guarantees, anticipation of fears, worries and anxieties, listening and reception processes, human contact, protection and security of personal data (confidentiality and privacy), interdisciplinary and cross-disciplinary collaboration, number and application of ethical rules, clarification of regulations relating to AI;

– mitigated: organizational silos, the role of the human being in tedious and arduous tasks, free will (autonomy), individualization of risk and responsibilities, isolation of certain people, laws and regulations that are too restrictive and not adapted to the evolution of industry and the economy;

– elaborated: tutorials on data and AI usage, cultural transformation around Ethics by Evolution, a way to distinguish human from AI, new positions such as a CDE, guarantor of digital ethics especially for applications related to AI, product experiments;

– deleted: management platforms, some business lines like customer service;

– developed: the right to obtain information about the AI system. The ideal would be that they are provided without making any request. It would be possible, for example, to create an obligation for the creators of AI systems to communicate that is understandable by all. A first level of communication could consist of the publication of a declaration of the intentions to use the system in a sincere, fair and complete manner. For AI systems that involve the creation of a "user profile", a second level of communication could consist of publishing the information necessary to understand how the AI system works within this personal space. Finally, a third degree of communication could require, in the presence of personal data, to identify an easily reachable team responsible for the operation of an AI system.

We are finding that monetizing our data, or arranging it into information sectors, is causing more suspicion and leads us to question which governance would be better employed. We are witnessing a new algorithmic governmentality that fits into the global context of capitalism, and which thus favors the individualization of service offers or consumption. In this way, the user is certainly placed at the heart of the device, but no longer given the opportunity to express one's intentions, desires, motivations and preferences, which are automatically inferred by digital devices.

Consequently, each model and algorithm is associated with hypotheses that need to be (re)specified and intrinsic evolutionary parameters that need to be refined, which bring to light data collected under very specific conditions.

In addition, in a world where data exploitation has become a major economic issue, companies are forced to integrate strong analytical models to extract decisive information and target their commercial offers. As a result, decision-making is mainly driven by data, and no longer by the intuition of the players in their market. Such a shift in business culture requires a rigorous and coherent management around Big Data and AI that use, process or analyze it. This data management concerns all the layers of the company, because this analytical modeling is present in all the services[19] of the structure. A relevant management of these important data volumes is an indispensable condition in order to enable companies to implement rapid actions to improve their performance and efficiency. This requires organizations to employ analytical models that are flexible, scalable and easy to adjust and maintain.

Consequently, it is essential to rely on management tools that allow a large number of people to quickly understand AI and the Big Data associated with it within the company. This is why the challenge of tomorrow lies on the fact that every person can understand the meaning of the processed data.

---

19 The range of services includes operational services, finance, quality control, marketing and communication services, and technical services.

A good management of digital data used by an AI must define the rules, devices, limits and organization to ensure the management of data assets in line with the strategy and requirements of the company. All Big Data management actions are at the service of various purposes, but all are essential and must be articulated between them:

– produce: define and launch the processes, services and equipment that generate data, enabling the company's programs to be met;

– collect: to collect the data available, in a structured or unstructured form, without necessarily knowing *a priori* what they will be used for in the future;

– store: retrieving and storing a huge and growing volume of data, trying to organize and streamline this stock to facilitate further data processing;

– study: to carry out in-depth studies in order to internally or externally enhance data;

– value: exploiting the potential of the data, internally or through partnership or commercial relationships.

After a phase of emergence and discovery, AI associated with Big Data is now relatively accessible. The necessary change within a company is now cultural in order to take all the potential benefits of these "mega-data".

In addition, the management of AI requires a high level of transparency and respect for the rights of individuals with regard to their data. In particular, it is a question of informing the customer about:

– the presence of a personal data collection process;

– the nature of the data processed: personal (name, address, professional situation, etc.), sensitive (state of health, religious, political, sexual orientation, etc.) and other types of information essential for tracking navigation and personalization (cookies, logs, etc.);

– the purpose for which these data are used;

– the possibility to access the stored customer data and/or to refuse this collection or support device.

Big Data management contributes to the valuation of the information that comes out of it by facilitating its understanding and, therefore, its uses. The added value of this management is linked to the time it saves in the quest for the "right" information for a good use, and in the ability to save time for the company in the search for understanding Big Data and the meaning associated with consistent information.

> COOKIES.–
>
> A cookie is a small text file in alphanumeric format deposited on the hard disk of the Internet user by the server of the site visited or by a third-party server (advertising network, web analytics service, etc.). According to the CNIL, a cookie is a sequence of information, generally small in size and identified by a name, which can be transmitted to your browser by a website to which you connect. Your Web browser will keep it for a certain period of time, and will send it back to the Web server each time you reconnect to it.

> LOG.–
>
> In computer science, logging represents the concept of history of events designating the sequential recording in a file or database of all events affecting a particular process (application, computer network activity, etc.). Usually dated and classified in chronological order, logs allow step-by-step analysis of the internal activity of the process and its interactions with its environment.

Finally, managers of Big Data and processing algorithms have to ask themselves a whole series of questions such as:

– What are the major data to manage? What are their sources? Who are the producers? Is it necessary to go back over the history?

– How to eliminate the "noise" that disturbs the readability and understanding of the data?

– Is it necessary to centralize data in AI?

– How can we obtain a good data traceability? What is the lifecycle of data within AI? What are the different transformations performed? What are the intermediate data produced?

– At what level[20] of description are the data?

---

20 Model level/flow level, business model level/application model level, document/file level, etc.

– What organization should be put in place to best manage this Big Data?

– Should the data be opened externally?

– How the data be connected between them?

– Who are the actors and producers of these AI-based treatments?

– How can the effort-reward balance be managed? Which AI should manage them?

– Isn't the neutrality of algorithms a myth, and their manipulation a reality? Wouldn't it be necessary to develop the control of the results produced by the latter, by creating a profession of "algorithmists" composed of experts in charge of carrying out controls in order to verify the reliability and validity of the algorithms?

– Have all the regulations, specific to the nature of the data processed, been considered?

– What is the ownership of this data? What are the applicable licenses?

Such questions cannot be answered without a perfect knowledge and understanding of the environmental digital ecosystem in which AI and the associated digital data are living.

Thus, it is essential to establish rules of governance for digital projects, particularly in AI, following the example of CERNA's recommendations on Machine Learning (see Appendix 5). There are three rules of governance for each environmental parameter associated with digital:

– Structural:

- implement processes to attest the loyalty of algorithms, i.e. their conformity and ethical acceptability. This can result in the establishment of an independent audit agency for AI systems to identify undesirable consequences and unjustified bias. An independent authority will be entrusted with the task of *ex post* control, auditing and ruling on algorithmic processing for which a presumption of non-compliance with ethical principles has been registered with the Defender of Rights and Freedoms. The independent authority would be in charge of setting the conditions, and issuing and withdrawing these labels. Establishing a reporting procedure where any failure to comply with the ethical charter would lead to the loss of the label. This label would reinforce the attractiveness of the digital domain and ethics;

- develop a new EU surveillance agency responsible for protecting public welfare through the scientific evaluation and monitoring of AI products, software, systems, or services. This may be similar, for example, to the European Medicines Agency;

- evaluate the reliability rate and the quality of an information in order to optimize its use.

– Technological:

- develop agreed measures for the reliability of AI products and services that need to be undertaken: either by a new organization or by an appropriate existing organization. These metrics would serve as the basis for a user-driven benchmarking system for all commercially available AI offerings;

- write a mandatory technical data sheet validated by an audit. Any algorithmic processing will be delivered with a data sheet and a mandatory audit report prior to its use and carried out by a structure approved by the CNIL. The purpose of this audit will be to certify the conformity of the processing to its technical data sheet, which will specify a technical documentation including in particular its reliability limits, an argument justifying the respect of the above-mentioned legal principles, the purpose of the processing as well as an intelligible user guide for the executing personnel for whom it is intended;

- verify that AI does not possess in an autonomous way the capacity to injure, harm, destroy or deceive human beings. An AI-based deadly arms race based on AI should be avoided.

– Strategic:

- have a data management policy based on a limited and known set of principles. This change management program must be carried out over the long term;

- develop an "AI studio" to share open-source AI bricks and various skills (agile team of developers, data specialists, researchers, students) in order to accelerate real projects brought by local companies. Ideally, economic prosperity and the power released by AI should be fully shared, especially for the benefit and respect of all humanity through established social and civic processes (common good);

- implement a data reduction strategy, where only data with a defined value are kept.

– Methodological:

- not to make decisions on obsolete information;

- assess which decision-making tasks and functionalities should not be delegated to AI systems, through the use of participatory mechanisms to ensure alignment with societal values and understanding of public opinion. This assessment should consider existing legislation and be supported by an ongoing dialogue among all stakeholders (including government, industry and civil society) to discuss the impact of AI on society;

- apply governance methods to business priority data.

– Organizational:

- develop a framework to improve the explainability of AI systems that enable socially meaningful decisions. The ability for individuals to obtain a factual, direct and clear explanation of the decision-making process, especially in the case of undesirable consequences. This is likely to require the development of sector-specific frameworks, and professional associations should be involved in this process, along with experts in science, law and ethics;

- set up steering and ethics committees for the management and development of data within the organization;

- to set up "information protection officers".

– Regulatory:

- develop legal instruments and contractual models to lay the foundations for a fluid and enriching human-machine collaboration in the working environment;

- assess whether existing regulations are sufficiently ethically based to provide a legislative framework capable of keeping pace with technological developments. This may include a set of key principles applicable to emergency situations and/or unforeseen problems;

- assess the capacity of existing institutions, such as national civil courts, to redress errors or harm caused by AI systems. This assessment should evaluate the presence of a sustainable, agreed-upon foundation for liability from the outset of design to reduce negligence and conflict.

– Relational:

- make AI understandable by reinforcing existing rights and organizing mediation with users;

- not neglect the human (behavior and education) and organizational dimensions;

- develop a European observatory for AI. The observatory would have the mission to monitor developments, provide a forum for debate and consensus, provide a repository for AI literature and software (including concepts and links to literature), and issue step-by-step recommendations and guidelines for action.

– Cultural:

- ensure that AI designers and decision-makers remain socially responsible, sensitive, and directly engaged with the potential influences of AI technologies on society as a whole;

- not retain information longer than necessary;

- set up within the structure a CDE that is responsible for digital ethics.

3.9. Four key steps for an AI project

3.9.1. *Step 1: determine the project objective*

Objective: As Franck Goron (Senior Data, AI Strategy, and Business Program Manager) notes, any AI implementation project starts with a question: "If you had a magic wand, what would you like to do?" What would you try to predict? To understand? To streamline or optimize? What difficulty do you want to overcome? This phase of identifying the final business objective for the development of artificial intelligence within the company is strategic; it determines the next step in the process, namely the system and technologies to be implemented, the teams involved and the data sets needed and that can be mobilized.

Actors: This reflection can be carried out by the general management, but it can also be carried out at the level of each business line, like a production manager who identifies problems that he or she would like to solve in the production chain.

3.9.2. *Step 2: collect and prepare relevant data*

Objective: Once the project objective has been determined, it is time to turn to data. Data are essential to any artificial intelligence project, because it is the data that will feed the algorithms and allow the system to deliver a result. There are many requirements to guarantee the quality of the data used: Are they reliable? Are they up to date? Are they sufficiently diversified? Are they appropriate to the context?

Once collected, the data must be prepared: removal of duplicates, missing data questions, standardization, scaling issues, sample quality and distribution, and so on. This work is especially important when the data were distributed in different silos. It can be so important that it can sometimes consume more than half of the project time.

However, there are often false preconceptions about the preliminary work required to prepare the data: no, it is not necessary that the data are already structured and ready to be assembled. The computing power of today's computers makes them capable of proposing a structure on their own.

Players: An internal CDO or an external partner can assist you in this crucial process of collecting and cleaning usable data. If the data are not directly available, it is always possible to improve the data collection process internally or even to acquire data from data resellers. Microsoft is also currently testing a system, the knowledge graph, which will allow external data and data from different systems to be collected and aggregated and then used as a classic recommendation graph.

> KNOWLEDGE GRAPH.–
>
> The knowledge graph is a knowledge base used by Google to compile the results of its search engine with semantic information from various sources.

3.9.3. *Step 3: classify data and choose tools*

Objective: The collected data will then have to be prepared, classified and categorized in order to define the formula of the algorithm that will process these data.

At this stage, a cloud infrastructure is required to provide the computing capabilities needed to identify the different examples, or "patterns".

Actors: Internally or externally, the data scientist and data engineer prepare, then control and adjust the algorithm formula to achieve the desired objective.

3.9.4. *Step 4: produce the model*

Objective: Once this pre-processing has been carried out, and certain "patterns" identified, data scientists must choose the most appropriate algorithm. For this purpose, they benefit from libraries of preconceived algorithms that they will have to train in order for the system to perform the desired function: optimizing, predicting, etc. To train this model, data scientists must adjust its various parameters and train it through recurrent exercises. The goal of these numerous tests is to gradually refine the model in order to achieve a very small margin of error. Once the right model has been found, it is then possible to feed the system with new, totally unknown data to ensure that the results are in line with what is expected.

Once set up, it is essential to maintain your algorithm, or risk seeing your results falsify after a certain period of time. As reality and the data evolve, adjustments may be necessary. Indeed, companies must keep a close eye on the maintenance of their machine learning algorithms. In addition to maintaining the people involved, it is also essential to have a common platform and well-documented code.

Actors: This step requires the help of a data scientist – who may be in-house or with partners – and then a developer to integrate the finalized model into the customer's solutions (applications, etc.).

In conclusion, these main techniques allow AI system designers to better consider the respect of fundamental rights in the development of these systems. However, even if they are applied to the letter, they are not as such sufficient to fully and effectively guarantee adherence. Indeed, without calling into question the scientific integrity of the designers of AI systems, the respect of fundamental rights has always required the implementation of control mechanisms: scientific integrity is guaranteed by ethical responsibility. In this sense, all initiatives of ethical regulation of AI applications refer to an obligation of accountability. Hence, the importance of the implementation of ethical compliance control processes.

3.10. Algorithmic responsibility

Philosophical reflections on responsibility are traditionally oriented toward the human components of moral action. Generally speaking, when it comes to designating moral responsibility, one presupposes human actors who carry out acts with immediate and well-defined consequences. However, in a society that is becoming increasingly digitalized and where technology takes a pre-eminent place, human activity cannot be properly understood without reference to technological artifacts, which has the consequence of bringing certain complications to the assignment of moral responsibility. Thus, the emergence of AI, machines and robotics as new agents inevitably disrupt the attribution of an individual or even collective responsibility. This question on the attribution of a responsibility is crucial. Indeed, a democratic society progresses only by appealing more to the ethics and responsibility of its citizens. Thus, the way in which a person reflects on the existence, in the sense of things, of entering into a contingency will contribute to making him or her enter into a reflexive ethics based on values and goals. This acceptance of presence and openness to the inter-human or human-machine relationship allows for the awareness of the value of things around us to open up to responsibility and interdependence.

In the context of technology, ethics deals with acts, actions that have an incomparable causal reach toward the future and that are accompanied by a predictive knowledge that, however incomplete, also goes beyond what we have known in the past. To this must be added the order of magnitude of long-term actions and very often also their irreversibility. All this places responsibility at the center of ethics, including the horizons of space and time that correspond to those of the four ethical principles seen above.

According to philosopher Emmanuel Levinas, responsibility is the fundamental structure of subjectivity; it is what makes the human being a subject. Here too, responsibility is not reciprocal: one is responsible for others, for what is not one's own doing. With the arrival of autonomous machines, the boundary between objectivity and subjectivity is increasingly ambiguous. This calls into question the initial points of reference, where from now on robots also become a subject and an agent in their own right. Some people even propose creating a specific legal status for NICTs. This is the case of lawyer Alain Bensoussan, who advocates and campaigns to endow robots with rights and duties. Under these conditions, the category of the "legal personality" of a company under the law could be extended to include non-human actors such as companies (referred to as "legal persons"). This information allows us to reflect on the issue of algorithmic liability and to apply our Ethics by Evolution approach within structures and organizations.

At first glance, the idea may seem completely abusive, provocative and even disturbing. Is it the first step before thinking that the machine is equal to humans, before surpassing them? In our opinion, we should not look at the problem from this angle, but rather say that the attribution of an established legal status for autonomous devices will make it possible to shed light on the perimeter of the responsibilities of each protagonist. The designation of a legal profile makes it possible to clarify the status of the living and/or artificial agent and to associate rights and also duties to them. Given that the algorithmic responsibility is revealed in the duties of a digital actor, this step appears decisive for thinking about tomorrow's world! This approach would have the advantage of being able to propose a plausible alternative to another idea, which is that behind every machine, there is a human being or a unique "legal personality", which is not conceivable. Indeed, this vision omits the fact that complex devices are mostly the product of a multitude of authors, and that human beings and machines do not operate in separate environments. Exchanges, mediations and human-machine interfaces are so intimately associated that the very notion of a "sovereign human agent" is highly unlikely.

Therefore, such a reflection leads us to wonder about the importance for citizens to be, from a legal point of view, the true owners of their personal data. Indeed, we can think such a status could lead individuals to pay more attention to the importance of digital data, since they can now consider it as an intangible asset associated with a specific value. In this context, individuals will become more autonomous and actors (Empowerment) in the digital economy, and consequently more responsible in their decisions made about their personal data.

It should be noted that at present, from a legal point of view, European citizens do not own their data. Since digital data are not considered as an intangible good, but rather as an extension of the individual, the individual only has a subjective right[21] or a right of use. It is highly likely that this legal situation will evolve over the next few years in order to be consistent with the digital reality of the market.

Moreover, the notion of precaution[22] is increasingly often invoked in political and legal decisions where a medium and long-term risk for society as a whole must be assessed in a context fraught with uncertainty. It therefore also involves real philosophical problems, of an epistemological as well as ethical nature. Elevated to the rank of "rule", precaution takes on a higher dimension here. It is then a matter of making it a general theoretical rule that guides conduct. This brief definition contains all the ingredients of the problem such as risk, uncertainty of knowledge, legitimacy and, therefore, responsibility.

Thus, precaution is the act of prudence being carried out in a specific case. This means that, for antiquity, prudence is a general virtue of which precaution would be the effective consequence. Thus, prudence is the calculating and farsighted part of moral virtue and thus of wisdom. Note that Ciceronian *prudentia* is the direct heir to Aristotle's *phronesis* in his *Nicomachean Ethics*. In the face of serious and irreversible, but potential risks, the absence of scientific certainty should not delay the adoption of measures that would have been considered legitimate if such certainty had been acquired.

Moreover, most authors situate the notion of precaution in the direct filiation of an ethical reflection, started by the German philosopher Hans Jonas, founder of the concept of responsibility, who wondered about the evolution of our modes of action within technological civilization. Hans

21 Subjective right is a legal prerogative attributed to a person by law to govern his or her relations as a member of society, which he or she may avail himself or herself of in his or her own interest. This right may be of different kinds: the right to freedom of expression, the right to claim a debt, the right to own property, etc.

22 From a literal point of view, precaution comes from the Latin word *praecavere* (taking care), which implies a danger that has just appeared.

Jonas' heuristic of fear is, therefore, at the origin of a possible awareness of the long-term effects of our present technical and scientific innovations. The author advocated for an "ethical watch" that must support a "technological watch": the combination of the two founding a true "legal and jurisprudential watch". According to him, it is necessary to "take precautions" in the face of the risks of the future, then insisting on the need to balance reason and will in order not to slide into sterile fear. The ethics of the future renews with Cicero's *prudentia* and Aristotle's *phronesis*; the long term is the foundation of the short term and the common interest is the bearer of individual interest. It is, therefore, a question of combining precaution and prudence. Its general meaning designates social behavior confronted with the uncertain nature of scientific and technical evolution.

From now on, the ethics of beneficence, foresight and prevention will be succeeded by the ethics of precaution – and therefore of responsibility – which introduces the obligation of foreseeing the inherent risks of digital professionals by considering all their foreseeable consequences. This leads to an obvious conclusion of an ethic of the future, considering that what is possible is not necessarily "ethically acceptable or risk-free", the very principle of the evaluation of our long-term responsible acts, where it is not only a problem for the medical profession, but also for the entire community, to think about the nature of the responsibilities attributable to the actors of the digital world.

It is perhaps in this sense that the importance of the precautionary principle should be considered from the very conception of an NICT, which would ultimately have fewer rigid and fixed legal implications than a flexible and evolving ethical framework. Indeed, in the face of rapid technological developments, one may legitimately be concerned about the degree of reactivity, or even inertia, of legal rules, their capacity for rapid adaptability, and their ability to subject algorithmic processing to the regulatory and governance mechanisms that are required. This is why we propose an approach centered on an evolutionary ethics of the digital world, namely Ethics by Evolution or Responsibility by Evolution.

Even more significantly in terms of legal liability, the decision-making of a fully automated system also depends on the ability of the device to self-educate, i.e. the ability of the algorithms to learn and adjust the sequences that encode them, independently of any human control. Finally, the solutions to be provided to the questions raised about digital technology do not rely solely on the adoption of binding legal frameworks. It would, therefore, be essential to integrate an instrument, in the form of a declaration including all the ethical and legal aspects, in order to anticipate the future challenges surrounding NICTs. In this governance to be built, the place of public and private institutions (as trusted third parties) will have to be clarified.

In addition, some experts consider that the companies concerned by digitalization should first adapt the concept of "corporate social responsibility" (CSR) to the context of the launch of AI systems. The idea of developing the traditional concept into a concept of "digital social responsibility" seems compatible with the definition adopted by ISO[23]. Indeed, according to this definition, social responsibility is synonymous with "the responsibility of an organization for the impacts of its decisions and activities on society and the environment, resulting in ethical and transparent behavior that contributes to sustainable development, including the health and well-being of society". For its part, the CIB[24] has long held that "in view of our responsibilities to one another, it is important to recognize that the achievement of social responsibility objectives implies ... acceptance of the responsibility to minimize or eradicate risks to health and well-being. Social responsibility is a principle that defines and enhances our common humanity and our mutual commitment to improving the health of individuals and communities. Since ethical behaviors are by nature constantly evolving as they reflect the values that a society chooses to defend at a given time, it is, therefore, possible to consider that the concept of CSR is itself in the process of evolving according to the context in which it is applied. Thus, the new concept of "digital social responsibility" applied to companies offering innovative perspectives in the health sector should push them to invest more in the regulation of AI systems, or even to engage in voluntary certification schemes to attest to their ethical and responsible character. Indeed, as Gruson (2019) pointed out, "the implementation of an ethical and responsible approach to the launch of AI systems in a given sector of activity can be considered as a lever in its own right by which a socio-economic actor exercises its social responsibility" (Gruson 2018).

Moreover, the great ideologies associated with the industrial revolutions made our intimate and profound relationship with the planet abstract. Even a

---

23 ISO 26000 Standard [Online]. Available at: https://www.iso.org/iso-26000-social-responsibility.html.

24 CIB (2010). Rapport sur la responsabilité sociale et la santé, paragraph 100.

decade ago, we could still hope to mitigate environmental crises and their economic and social repercussions. But the reality is clear; we have never emitted as much greenhouse gas as we have since the Kyoto Protocol and the COP[25] (Conference of the Parties). Biodiversity is collapsing at a dizzying rate and the stocks of abiotic resources are almost all exhausted. The relationship between humanity and the planet is at a pivotal moment in its existence, as digital technology can both increase our ecological footprint and provide us with opportunities to reduce it and accelerate the environmental transition. In this context, the digital revolution and environmental transition are the two great transformative forces of the 21st century. As a result, this digital transformation offers new and innovative ways and practices in all sectors of life and can, under these conditions, represent a solution to accelerate the ecological transition, such as the integration of smart power grids, teleworking and teleconferencing, intelligent mobility, dematerialization, measurement and urban environmental monitoring (UEM[26]), energy-efficient smart buildings that integrate consumer and producer equipment, and energy storage equipment into their internal management. However, if the environmental challenges associated with the digitization of the world are not taken into account or are poorly addressed, we are exposing ourselves to a significant increase in our ecological footprint and the consequences that result, including the degradation of ecosystems and biodiversity (eutrophication of water with the proliferation of algae), asphyxiation of aquatic ecosystems), increased climate change related to global warming, increased greenhouse gases, air pollution, ocean acidification, depletion of the ozone layer, and depletion of natural resources and the circular economy.

Under these conditions, it seems essential to adopt eco-responsible behavior in favor of the planet. In order to do so, we must put in place certain actions and take decisions that are assumed to be in favor of the planet:

– a decrease in the direct negative environmental impacts of digital technology;

25 Conference of the Parties to the United Nations Framework Convention on Climate Change.

26 UEM consists of collecting and analyzing environmental and urban information with the aim of improving the performance of urban services, ensuring the monitoring of the environmental criteria of local authorities, and developing new services for local authorities, businesses and citizens.

– the development of common action strategies, both in terms of impact reduction and solutions that encourage or even accelerate the ecological transition;

– improved natural resource management, information flow management and human assistance;

– the structuring and emergence of a common culture between digital and environmental players in order to facilitate the articulation between Green IT[27] approaches to move toward responsible eco-design of digital goods or services (via methods, tools and good practices that share eco-responsible and sustainable development).

We then move toward the implementation of a methodology centered on Ecology by Design integrating:

– proactive (not reactive), preventive (not curative);

– implicit (default) consideration of ecological issues;

– ecology as an integral part of the design process;

– ecology from end to end – throughout the lifecycle;

– waste reduction (4Rs: review, reduce, reuse, recycle);

– visibility and transparency to ensure openness (for example, every digital application should have a "notebook" that accompanies it from conception to the end of its life: this information makes it much easier to repair, customize, resell and recycle);

– extending the active life of equipment by encouraging their reuse and postponing as far as possible the inevitable recycling stage: combating the obesity phenomenon, which is one of the main triggers of equipment obsolescence and a major factor in the oversizing of infrastructures;

– respect for users and their needs (sobriety, frugality, relocation of practices, etc.).

In addition, companies seeking to assess the legal risk associated with the implementation of AI must adopt a transversal and global approach to

27 The four main actions of Green IT are to reduce the ecological footprint of cloud computing, use digital technology to better design green policies, support digital innovation in favor of ecology and mobilize the potential of data to serve the ecological transition.

accountability, assessing their risk according to the different levels of ethics we have seen above:

– responsibility for the intent, effect and impact of an AI system (ethics of practice/ethics): use value. Enterprises must be certain and convinced that their use of AI is ethically justifiable and legally acceptable. To do this, they must clearly and precisely define the problem that AI seeks to solve and ensure that the algorithmic system works as originally intended. The absence of a human decision-maker in a device should not mean that liability for wrongful acts (such as discrimination against a person) resulting from a decision made by an AI is avoided. Furthermore, firms must also take a macroeconometric approach to determine whether the intent of AI they use is consistent with good corporate behavior. For example, decisions by an AI that affect the rights of individuals may negatively affect the reputation of the firm even if legal obligations are not violated;

– responsibility for the performance and actions of the algorithm (ethics of algorithms): management value. Firms must seek to prove that the algorithm's results have been achieved within reasonably acceptable parameters. This requires an examination of the development and construction of the algorithm itself, the data on which it was formed, and the testing of predictable results. Firms must, therefore, ensure that the AI decision-making system remains within the parameters set, because the more data that are associated with AI, the more the decision-making processes evolve over time. A careful audit on the security, bias, and discrimination risks seems to be crucial to determine the reliability, performance and fairness of the algorithm;

– responsibility resulting from the data used to form the algorithm (data ethics/ethics of systems: design value. Companies must have the guarantee, on the one hand, that their data are correct, honest and sufficient for the algorithm to make the right decision, and, on the other hand, that their systems for collecting, using and disclosing their data are credible and reliable with a view to the confidentiality of personal data. To do so, companies must maintain the security and integrity of personal information. In addition, algorithmic systems must be tested to ensure that the intended use does not inadvertently disclose personal or even sensitive information (model inversion)[28]. This test can also be a security measure to prevent a careless person from acquiring personal information through intentional misuse of the system.

28 Model inversion is an AI risk that occurs when a user has certain data about a person, but can then establish other information about that person by observing the result of the algorithm.

In the end, the question of the liability of a company using an AI is as vast as the potential use cases are numerous. In the majority of cases, liability around AI will be simple, without the need to test the limits of established liability frameworks. However, some complex algorithmic processes (even with artificial and self-evolving consciousness in the case of "strong" AI[29]) will require careful thought and further legal analysis. Companies must also take into consideration the significant number of policies and regulations being developed around the world to design and implement guidelines on acceptable parameters for AI use.

Finally, the digital revolution imposes a framework of clear and coherent ethical and moral rules in relation to the set of common law rules with which they interact in order to restore the trust and meaning that citizens expect from these NICTs. These ethical recommendations cannot be applied without a manager dedicated to this. It is the role of the CDE to monitor the ethical impacts of algorithmic systems.

---

29 The topic of the supervision of "strong" AI with the capacity to be autonomous, evolutionary, and with an artificial consciousness, will feed our prospective reflections in Chapter 4.

4

Anticipation Around Artificial Consciousness

Understanding the mechanisms and processes of human consciousness is one of the greatest scientific challenges of the 21st century. We all know how to globally define a consciousness, yet no one today can claim to understand in depth the structuring of human consciousness. No one really knows how and why brain activity is associated with a subjective mental life lived in the first person.

The latest AI applications make it possible to simulate all human reasoning when applied to rational knowledge domains. Computerized systems, all systems with processors and memories, can now be equipped with the ability to generate intentional thought forms, to have wants and needs and to wrap human users in sets of procedures that they can no longer control that are beyond themselves. In a few decades, so-called "strong" examples of AI will be aware of themselves and their environments and will be able to deduce new rules from them, theoretically capable of adapting to any type of environment without the need to reprogram them. If all multi-agent systems are endowed with their own consciousness, then they will be able not only to respond to their own desires and fears, but also to represent the desires and fears of other agents. They will then be able to anticipate the reactions of the environment and other agents, and plan their actions accordingly. One can even imagine an artificial consciousness capable of creating itself a new consciousness about entirely new things, resulting from its own reflection! Nothing will prevent a computer from inventing a new religion that will cause new suffering for humanity. Thus, if a machine has a consciousness, it will have the choice to do what it was created for.

Therefore, the development and exploitation of artificial psychological systems endowed with intentional consciousnesses must necessarily pose the problem of the choice of their uses or the justified decision of their non-realization. It is essential for humans to think about this new intelligence so as not to be surprised by the effects and consequences it could generate. For this consciousness and these affects will be those of a machine and will, therefore, not be perfectly comparable and similar to those of the human species. Therefore, the ethical problem of the use of artificial consciousness must now be clearly raised.

Under these conditions, it becomes fundamental to enter the architecture and the bowels of human consciousness in order to think about designing and applying it in AI. After having sketched the structure and the dynamic mechanisms that constitute an artificial consciousness, the major finality of our research will be to associate and to apply our approach of Ethics by Evolution to the artificial consciousness, so that this one is ethical, moral and eco-responsible for humanity.

Contrary to existing theories, we defend the idea that consciousness is something that evolves and structures itself over time, in parallel with the cognitive and intellectual capacities. Our experience of reality is limited by what our senses allow and give us to perceive. Consciousness is, therefore, something that can be learned. It results from learning processes by which the human brain continuously learns to predict, understand and analyze the consequences of its own activity on its own functioning (internal loop), on the external world (action-perception loop) and on others (self-other loop). From this observation, we can say that consciousness emerges from the knowledge we have of ourselves, as well as the interactions we have with our body, our world and others.

How is the awareness of our consciousness of our own intentions modulated by the perception of the brain activity that produces those intentions? How can living or artificial matter have a subjective aspect, generate feelings, affect, qualitative sensations and produce knowledge of things? Is there a universal morality? If good and evil are only a private matter, how can we agree on the common good? In the name of what principles do we want political action to be ethical? Where would this morality of conscience come from and what would be its conditions of possibility? Would it be an abstract conceptual elaboration? Would it be based on specific values? Is it part of an ontological relation, based on various modalities of responsibility toward the being? Is it necessary to constitute a library of doctrines, and ethical and moral elements? How can the appropriate ethical context be determined? How can the user easily configure the ethics of his or her artificial agent? What ethical limits and obligations should be set for everyone? How can the writing of moral/ethical elements for non-computer scientists be facilitated?

To try to answer these questions, we start from the following premise: "everything in the universe is a question of information!" Humanity, physical and biological laws, space-time, and matter are governed by information. Information is, therefore, the original source and primitive matter of the Big Bang describing the origin and evolution of the Universe. Claude Shannon's theory of information appears to be at the heart of this debate associated with the neurosciences. Moreover, the word "consciousness" comes from the Latin term *conscientia* (*conscire-cum-scrire*: "knowledge", therefore shared "knowledge"). Consciousness is based on a shared knowledge with someone oscillating between the values of "trust" and "connivance", and it is also a clear knowledge that one has deep within oneself, as a faculty that human beings have to understand their own reality.

Therefore, from this statement, we can hypothesize that any system capable of integrating information, whether made of carbon or silicon, should generate states of consciousness. Thus, it seems relevant, and even coherent, that the systemic process at the origin of the creation of practical wisdom from information, and then from knowledge, is transposable to describe and understand the device for the creation of a non-biological artificial consciousness. Neoplatonic modeling centered on the information theory embedded in neuroscience would make it possible to measure the degree or even the index of consciousness and to express this phenomenon in purely mathematical terms.

4.1. Protean aspects of consciousness associated with intelligence

The appearance of human beings – and their intelligence – did not happen all at once, but as a result of a succession of small transformations over the millennia, sometimes in the right direction and then preserved by evolution, at other times useless or unfavorable and in such cases eliminated by it. For it is always in this way, by "trial and error", that evolution proceeds, with natural selection as the driving force behind the movement. Of course, the human being has not escaped this law and this is what explains the immense time that separates us from the first hominids.

The conscious mental state is revealed through some form of representation. These representations can be of very varied natures – from the most cognitively developed to the most primitive. The most frequent are determined by the perceptions of our senses, triggered by stimuli or signals coming from our body (sensations, touches, pain, etc.) or the external environment. We agree to distinguish two different natures of conscious experiences: on the one hand, a phenomenal aspect (close to impressions and sensations) and, on the other hand, a co-generative or access aspect (close to thought and action):

– *phenomenal consciousness* makes us aware of the effect that the experience of the surrounding world has on us through our senses, and also of the effect that we have on ourselves through the sensations of our body. It includes subjective consciousness or qualitative consciousness "for me" (object consciousness) and impersonal qualitative consciousness (object consciousness);

– *access or cognitive consciousness* gives us access to conscious behaviors and voluntary and intellectual activities in general. In other words, it translates intentionality directed toward cognitive functions in general, control, actions, etc.

We support the thesis that cognitive consciousness does not and cannot have a separate existence from phenomenal consciousness. It enters into the structure of the direct object consciousness – where the object can be internal (thought object) or external (perception object) – of which it is the cognitive aspect. The representation of the object gives rise to the consciousness of the (represented) object.

In addition to the two main meanings seen above, the concept of consciousness has many meanings, characteristics or manifestations that one can try to distinguish, although occasionally these differences are in particular differences of degree:

– subjectivity;

– the relationship with the self;

– knowledge and experience;

– memory (procedural [habits and characters], semantic and episodic);

– temporality;

– the affect (sensations and emotions);

– free will and availability;

– selectivity;

– synthesis and unity;

– attention;

– planning, imagination and creativity;

– intentionality.

With regard to the criteria that make up consciousness, the latter is in essence strongly linked to intelligence, which includes the ability to think rationally, coherently, with an intention, and which is itself multisectoral. Intelligence subordinates the decision-making process to values and ethical considerations. It enables people to form relationships, to react emotionally and, finally, to make mistakes and learn from them, which is an essential prerequisite for creativity and innovation. Thus, consciousness is the ability to evolve one's own reasoning and to modify it in order to improve it. Thus, from a single consciousness, it would become possible to obtain an intelligence capable of recursively improving itself (the "new" thought being in turn evolved and then improved), eventually improving the evolution process itself.

Moreover, the expression of consciousness fluctuates according to the situation:

– with regard to the merits, it represents the sense of the ideal and moral values;

– with regard to the rule, it expresses the sense of duty, obligation and responsibility;

– with regard to the value judgment, it is the power to assess conduct according to the criterion of the license and the prohibited, ratified by affectivity;

– the sense of surpassing oneself gives the possibility to transcend oneself.

It thus makes it possible to perceive the gap that exists between what is and what should be, between the real and the ideal, the existence and the value, the fact and the moral rule or the law.

The conscious being lives at the same time as the actuality of one's experience while being nobody of one's world. Consciousness is not a simple function of the being, but is also its organization. To be conscious is to be aware of a lived experience, of a field of experience, of an unimpeachable experience of the subject who lives it.

The consciousness of the individual is shaped by his or her social relationships: it is shaped by our relationships with others, and individual consciousness is not the source of morality. The consciousness of the individual is not only the reflection of society, but also the place of our interiority; it is subjective (singular) and specific to each one. The consciousness of the individual is only the reflection of society; our consciousness is determined by society and it is the reflection of our social class.

Since the work of Antonio Damasio, we know that emotions play a major role in rational behavior. It then appears that everything that is intelligence now seems to become cognitive. The different forms of intelligence are variants of the same cognitive capacities. In the end, action, perception and cognition are inseparable from each other, and are intertwined with the different manifestations and expressions of consciousness (see Box 4.1).

Intrapersonal

Ability to understand oneself, to understand one's emotions and intuitions.

Verbal/linguistic

The ability to express oneself in a way that is understood by all.

Ability to put words to emotions and feelings.

Corporal/gestural

Ability to coordinate body and mind.

Ability to express oneself with the body.

Interpersonal

Ability to feel other people's emotions.

Ability to study and interpret the behavior of others.

Existential/ethical

Competence to question oneself about life and how it works.

Ability to philosophize and be sensitive to spirituality.

Logico-mathematics

Ability to analyze numbers, calculate, measure, solve problems and logical–technical tests.

Musical

Increased sensitivity to sounds, tones and rhythms.

Ability to elaborate music.

Naturalist

Understanding of nature, the environmental ecosystem and animals.

Ability to interpret animal and plant life.

Visual/spatial

Ability to visualize the world in 3D, enhanced sense of orientation.

Box 4.1. *Different forms of intelligence*

4.2. Structuring of consciousness

As we have just seen, the evolution of consciousness cannot be dissociated from the intellectual development of the organism. Consciousness is structured with the development of semantic skills, the acquisition of language and symbolic references, spatial perception, etc.; in other words, all the dimensions that make up intelligence, and therefore part of cognition. Consciousness is not an object, but a process. It is closely related to subjective experience – what I feel when I perceive a color, recognize a musical note, or uncover a pain or, on the contrary, a pleasure. These impressions or sub-objective experiences (unique and singular) are called *qualia*.

From then on, to question one's condition as a "subject" is first of all to put oneself in relation to one's double, namely the "object". The space of one's identity realized as "subjectivity" is revealed in the gap between subject and object. Finally, questioning the nature of AI as a subject amounts to asking to what extent can we characterize AI as being endowed with consciousness? If we take up again the work on this subject, we notice that we can decline it according to three fundamental strata and levels:

– *self-consciousness*, as the consciousness that consciousness takes of itself and in relation to others;

– *consciousness of the environment and the object*, as knowledge of a knowledge or perception of a perception;

– *moral conscience*, as a moral judgment, giving meaning and significance that the subject carries on oneself.

These levels of consciousness can be studied separately, but are only components. The consciousness of the thinking subject is what builds on and unifies these phenomena. But at the same time, this functioning is fed and altered by all the unconscious cognitive processing, and by the phenomena of the unconscious, which by definition are not part of the conscious field, but which must be taken into account to characterize its functioning. If childhood and adolescence last so long in humans, it is precisely because it takes a long time to educate the unconscious processes of the brain and to create, within the unconscious brain space, a form of control that can, in a more or less reliable way, operate according to conscious intentions and objectives (see Box 4.2).

Unconsciousness

C0: Routine actions, automatic gestures, face and word recognition. When the brain has decided to act even before, we become aware of the intention we have acted.

Consciousness of the world (or primary consciousness)

C1: Representation and perception – even simplified – of the outside world (environment, society, space) and reactions to it. This consciousness is associated with phenomena relating to the sociological, political, economic and cultural context as a device of representation (culture of a country, corporate culture, mass culture, etc.).

Self-consciousness (or reflexive/introspective consciousness)

C2: Consciousness of specific phenomena related to the concept of self. Internal perceptions (own body), aspects of its personality and its acts (identity of the self, cognitive and affective operations, propositional attitudes). Capacity of the subject to perceive oneself as the author of one's thoughts. Self-consciousness is built from episodes of subjective consciousness. It is the presence of the mind to itself in its representations, as reflexive knowledge of the subject who knows himself or herself to be perceiving. This level of consciousness is linked to the notion of reflexivity (capacity to represent oneself) and metacognition (capacity of introspection). Only hominids, dolphins and elephants would have access to this level of consciousness.

Moral consciousness (or ethical consciousness)

C3: Understanding and ownership by the entity of the ins and outs of its actions for the community and future generations. The deepest essence is something immutable and constant: a sense of duty and obligation, a sense of responsibility and dignity, respect for ethical rules and human values, a sense of the ideal and the right use (notions of good and evil) to be made of freedom, justice and peace. This consciousness is both innate and acquired. The common moral values associated with this state of consciousness are, on the one hand, collective and relative to our society, and, on the other hand, individual and relative to our family environment and to what we are deep within ourselves *through* our free will. Moral consciousness is therefore the faculty of the organism by which it takes itself as the object of its moral judgment.

Box 4.2. *Structuring consciousness*

From this observation, we can associate the different forms of intelligence with the different levels of consciousness. These links of correspondence are essential in our process of identifying and structuring an artificial consciousness (see Table 4.1).

Levels of consciousness	Forms of intelligence
Unconsciousness (C0)[1]	Verbal/linguistic
	Corporal/gestural
Consciousness of the world (C1)	Musical
	Logico-mathematical
	Visual/spatial
	Naturalist
Self-consciousness (C2)	Intrapersonal
	Interpersonal
Moral consciousness (C3)	Existential/ethical

Table 4.1. *Links between levels of consciousness and forms of intelligence*

On the other hand, consciousness is a more or less clear form of knowledge that an entity possesses of its states, its environment, its reflections, ideas, passions, emotions, moods, identity and, therefore, of itself. The etymology of the word seems to confirm this hypothesis, since *cum scientia* in Latin means "with knowledge". Is it possible for a machine to hold this knowledge? What are the limits to this knowledge? How is this knowledge manifested and formed? What is certain is that the way in which I become aware of myself is similar to a determined cognitive process. This dynamic process starts from the direct relationship with oneself (the "self-consciousness"), which goes through a phase of "self-recognition" through the recognition of others and action, and finally ends in a stage of self-expression (the "outside consciousness") as a commitment, an opening, an encounter with the outside world. This last stage is associated with almost objective forms of collective and social consciousness. It is thus the consciousness of what one should or should not do according to principles

1 It can be noted that the forms of intelligences (verbal/linguistic and corporal/gestural) inevitably come from a consciousness of the world (C1) before being assimilated and integrated into an unconscious automatism (C0).

that place respect for others and for oneself above personal interests alone. This corresponds to a pragmatic and not only theoretical consciousness. It is thus a knowledge on which a will is articulated. This will is by essence personal, autonomous, but nevertheless placed under the authority of a duty that implies a form of universality.

Thus, to the notion of moral conscience is irremediably associated that of responsibility. AI must therefore be aware of its actions, to "know what it does": first, to recognize oneself as the author of the action, then to be able to anticipate its consequences (for itself and for others). Responsibility is thus in a direct relation with the act or action, which supposes the intention, the lived experience, the relation to oneself and to others. For, finally consciousness is always associated with something, with an object. It is in movement, intentional, open to the outside world, and not only closed in on itself in a reflexive way. Under these conditions, consciousness is, therefore, based on knowledge, expertise and perception, but it is not reduced only to that. It corresponds to a design, a plan, a projection toward the outside world. It is thus characterized by intentionality and not only as an interiority closed in on itself. Moreover, to be conscious of something is to know that one knows, to be aware of the knowledge one has, and to perceive that one perceives it. Awareness can be translated as an act of turning the self and the mind back on itself; in other words, a knowledge or a thoughtful perception. From then on, the consciousness considers itself as an object. Thus, reflexivity can be translated as the essence and intrinsic structure of consciousness.

We conclude that this intentional element of action proves to be one of the decisive pillars of moral consciousness, and, therefore, one that can be judged. However, on the other hand, there is no act without commitment, without realization, without externalization, without liberation and, therefore, without a complex social and historical reality. In a sense, "becoming aware" would be a reconnection with reality, a truth of something. AI learns in the reality of the external world in which it evolves. These events of reality learn something and awaken the AI to something. Nothing is static, and everything is in motion, intertwined, articulated and interwoven.

Finally, consciousness interacts with temporality, on the one hand, by being the preservation of the past and, on the other hand, by being the movement toward the future and anticipation. It is, therefore, the bridge and the link between the past and the future. Thus, consciousness reveals itself through memory. Without memory, a consciousness is without consciousness of itself; in other words, an "unconscious" consciousness that would be unable to identify anything and therefore perpetually confronted with something new.

4.3. Neoplatonic systemic ethical modeling (Ψ, G, Φ) of an artificial consciousness

How can an action be chosen? On which criteria? All the morality is reflected in these questions. The moral problem is none other than the problem of the choice of action and its ethical value, because action constantly forces us to take a position between the permitted and the forbidden, good and evil. Any commitment calls into question a set of values, a philosophy of life. We might as well act with full knowledge of the facts, making lucid choices, assuming the responsibilities and risks of these choices, but to do so, we must be conscious and voluntary. True human action implies becoming aware of one's own goals, introducing a reflection on the moral aspect of the action, monitoring its realization. For the human being, there is the possibility of aiming at the sensible end best suited to the goal to be achieved and of choosing, among the possible acts, the most adequate one.

Overall, the notion of complexity is based on the main idea that a system composed of different parts forms a whole that is different from the sum of its parts. Complexity is today part of a real movement of thought that invites us to restore the intelligence of complexity in our cultures and ways of acting. It poses a key epistemological problem for knowledge and action. Attaching oneself to complexity means introducing a certain way of dealing with reality and defining a particular relationship to the object. This implies that the very organization of these components creates emergences. That is to say that it produces specific properties that cannot be deduced from the knowledge of each of the parts. The analysis alone of the articulations between the elements is no longer sufficient. From now on, it seems essential to develop new tools of reflection, allowing for a better understanding and anticipation of the mechanisms of recursive logics, feedback and the phenomena of relative autonomy that make up an organization turned toward the creation of practical wisdom resulting from consciousness.

The individual perceives "data", interprets "information", makes connections with other memorized "knowledge" and is then able to act with his or her "knowledge"; he or she thus acquires skills so that he or she is able to implement repeatedly translating "practical wisdom" (Ricœur 1990). As a result, this approach, which is both epistemological and ethical, is part of the fundamental model for understanding the human dimension of the use of information and communication technologies (ICTs). Our transdisciplinary study naturally invites an epistemological critique focusing mainly on "knowledge processes" that integrate Leonardo da Vinci's Neoplatonist thoughts of "doing to understand and understanding to do" (Valéry 1941, 1948). This is a real challenge for knowledge, from a theoretical and practical point of view. Thus, this process of elaborating practical wisdom from consciousness allows us to move from a state (A) of complex, disorganized and fuzzy knowledge to a state (Ω) of simple, structured and teleological knowledge.

In order to succeed in this transition to simplification, we must use the right representation and modeling, which consists of substituting a process description for a state description. This transformation is carried out through a Neoplatonic systemic ethical model (Ψ, G, Φ) that integrates both:

– the reflections of da Vinci[2], Vico[3], Valéry, Morin, up to Lemoigne on modeling: ethics (Ψ: Psi[4]), epistemic and anthropological (G: Gnosis[5]), pragmatics (Φ: Phi[6]);

---

2 According to Leonardo da Vinci, "modeling (the Disegno) is of such excellence that it not only shows the works of nature, but produces an infinitely more varied number of them. [...] It surpasses nature because the elementary forms of nature are limited, while the works that the eye demands from the hands of the human are unlimited," by F. Tinland. (ed.) (1993). *Systèmes naturels, systèmes artificiels*. Champs Vallon, Ceyzérieu. We become aware of the importance of a system of representation that contextualizes the design: we want to do what, in what, for what?

3 "The ingegnio (the ingenium) is that mental faculty that allows one to quickly, appropriately and happily connect separate things" (Le Moigne 1999). *Épistémologie constructive et science de l'organisation*, Economica, Paris.

4 This Greek letter is often used to designate psychology, psychiatry, and psychotherapy, globally the sciences of human thought. This symbol means "soul, psyche".

5 The term *gnosis* means knowledge in Latin.

6 This Greek letter determines the golden figure measuring 1.618. Since antiquity, this symbol represents the divine harmonious proportion. It forms the basis of structures in archaeology, art, anatomy, music and literature, i.e. the concrete and practical aspects of science and art.

– the construction of an ethical event (intentions, situations and actions);

– Nonaka's model of organizational learning that surrounds dialogue and practice (socialization, internalization, externalization and combination).

Systemic modeling makes it possible to develop relevant methods of decisions and legitimation by considering the ethical issues underlying the event under study. Under these conditions, modeling consciousness allows us to better understand the experiences of the relationships we have with the world, in other words to "transform our experiences into science with consciousness" according to Leonardo da Vinci. This involves considering both relational and cognitive interactions with, on the one hand, the methodology used to "learn how to do" and, on the other hand, teleology aimed at "understanding what to do". Consequently, an improvement in thinking tools necessarily leads to an improvement in decision-making.

As we have seen previously, we can introduce the structuring of consciousness; in other words, the Big Data's path toward info-ethics in a Neoplatonic systemic ethical modeling (Ψ, G, Φ). By its very nature, a complex system is a dynamic process integrating multiple interactions and feedbacks, within which take place mechanisms that are very difficult to predict and control, which the classical conception was unable to envisage (Morin 2005). This open system is structured by an environment, functions, actions, teleology and transformations. Under these conditions, ethics must mobilize intelligence in order to confront the complexity that surrounds the communication process of medical information via NICTs. Thus, the intrinsic strength of this modeling lies in its interactive, multi-dimensional and active character turned toward the meaning, knowledge and teleology of an event.

It is composed of three different dimensions that articulate and interact with each other:

– *environmental dimension (φ)*: real[7] world environment: consciousness world (C1);

7 The environment of the real is a dimension favorable to location (identification, localization, estimation, characterization, mapping and hierarchy) and preservation (acquisition, formalization and conservation). The environment of AI is composed of four distinct plans: structural and technological, strategic and methodological, organizational and regulatory, and relational and cultural.

– *informational dimension (g)*: self-consciousness (C2);

– *ethical dimension (ψ)*: info-ethics (Fessler and Grémy 2001): moral consciousness (C3).

We note that G and Φ belong to an objective set called "informational rationality", while Ψ comes from a subjective set called "cognitive rationality" (Fransman 1994; Roth 2004). These rationalities are the subject of links between them in order to make possible the interactions between "acting" and "thinking".

Within an artificial consciousness, this informational rationality is made up of two loops of algorithms that work in parallel: one that allows the machine to understand its environment and one that allows it to observe itself in the process of functioning and adjusting. This system is made up of many small, very strongly connected elements and wonders how representational forms of the level of sensitivity of corporeality are generated and especially representing symbolic evolutions of things in the world at very high linguistic and conceptual levels. Regarding cognitive rationality (ethical dimension (Ψ)), it can be translated as a center of synthesis, organization and transformation (reasoning and deliberation) that starts from ontologies of knowledge about everything that we know today to represent cognitively and which wonders how to define hierarchies of systems that express all the uses of this knowledge from any points. These cognitive agents have internal memory and cognitive abilities and an internal state. It is a multisystem device fed by structurally morphological and semantic elements (representing and deliberating) from the environmental (Φ) and informative (G) dimension.

In order to study the architecture of an artificial consciousness, we decided to rely on an analytical framework composed of three reading axes: actions, situations and intentions that form the event. Intention[8] is associated with value. Therefore, as a consequence of our ability to intend ideal entities, some of them can become ideal models of our activity and can guide it, i.e. they can be presented as values. From this observation, the object of our research, in this case artificial consciousness, is rooted in a situation, in an intention that integrates values and in a determined social framework:

8 Intention represents the possibility of representing a state of affairs that is only ideal, that is not materially present, but that can be created by what is often called our symbolic "activity".

– *the intention* is the center of the discernment of good and evil, of the enlightenment of choice and of the judgment attached to moral values (affectivity);

– *the action* has for the role of acting with dynamism and will in order to accomplish a duty or an obligation;

– *the situation* that represents the context, the environmental parameters, the reality and the problem.

The model we propose shows the interactions between perception, cognition, language and emotions. Perception, emergence, recognition of concrete objects and thought are prior to the implementation of language. Therefore, they are considered as subsymbolic, and they result from the memorization of perceptual experiences.

For linguists, there is no thought without language, i.e. without words designating the objects of the world. Reflexive thinking is strongly linked to the consciousness of oneself, of one's body, to exist, and to be a living entity of the world with a history, memories, emotions and projects.

Thought is responsible for the content of our consciousness and for everything that happens in the world. Thinking is a reaction of the memory that contains knowledge, the result of experience that dates from the beginnings of the human being (phylogenesis) and from our birth (epigenesis).

Experience can be described by the following process: experience → knowledge → memory → thought → action → experience.

Finally, the thought that results from consciousness is a movement in time and space. Thought is memory, remembrance of past things. Thought is the activity of knowledge, knowledge that has been gathered over millions of years and stored as memory. In order to do so, artificial consciousness must be endowed with the following capacities: autonomy, beliefs, intentions in the form of a goal, an inner state describing its structure, actions, behavior, capacities, an outer state constituting its environment and communication (see Figure 4.1).

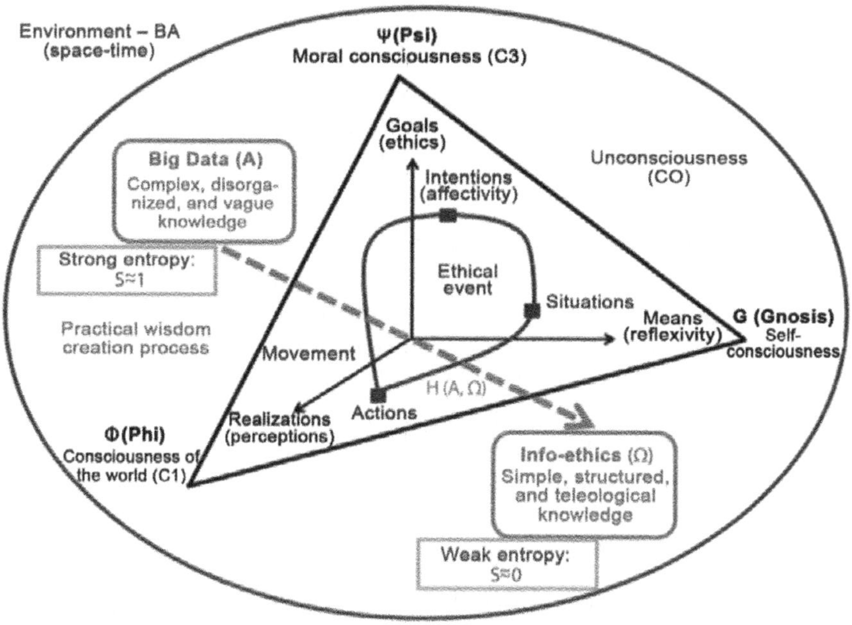

Figure 4.1. *Modeling an artificial consciousness through the Neoplatonic systemic ethical prism. For a color version of this figure, see www.iste.co.uk/beranger/responsibility.zip*

Finally, considering the differentiated aspects of the ethical dimension (Ψ) of information allows for the respective interaction with the environmental (Φ) and informative (G) dimensions. These three levels of ethics are an integral part of moral consciousness (C3) making reasoning, judgment, deliberation and thus moral decision-making possible (see Table 4.2):

– descriptive ethics (C3-1): this study refers to the "pragmatic" aspect and has as its mission the application of rules and practices. It deals with the means, procedures, processes and ways put in place to achieve the "carrying out" of objectives. It illustrates the methodological dimension (learning how to do it, in what context). This level of ethics interacts with the world consciousness (C1) included in the environmental dimension (Φ) of our Neoplatonic systemic ethical modeling (Ψ, G, Φ). These ethics represent the center of "acquisition" and "organization";

– normative ethics (C3-2): this space can be characterized by the "epistemic". It must concretize the goals established by the teleological dimension (to understand why to do it, for what project). This plan also refers to deontology, whose function is the regulation of practices. This level of ethics is in interaction with self-consciousness (C2) included in the informative dimension (G) of our Neoplatonic systemic ethical modeling (Ψ, G, Φ). These ethics are at the center of "transformation" and "interpretation";

– reflexive ethics (C3-3): this analysis is "reflexive" in nature. Its mission is to bring to any practice and any norm their "legitimization" (how to think, how to deliberate, what decision to take), hence a certain founding and synthesizing connotation. This space is oriented by values that serve as both normative and critical inputs. This level of ethics is in direct interaction with the descriptive and normative ethics composing the moral conscience (C3) included in the ethical dimension (Ψ) of our Neoplatonist systemic ethical modeling (Ψ, G, Φ). These ethics constitute the center of "synthesis" and "application".

<table>
<tr><th>Levels of ethics</th><th>Dimensions</th><th>Consciousness</th><th>Functions</th><th>Forms of intelligence</th></tr>
<tr><td rowspan="4">Descriptive ethics (C3-1)</td><td rowspan="4">Environmental (Φ)</td><td rowspan="4">Consciousness of the world (C1)</td><td rowspan="4">Acquisition and organization (pragmatic)</td><td>Musical</td></tr>
<tr><td>Logical-mathematical</td></tr>
<tr><td>Visual/spatial</td></tr>
<tr><td>Naturalist</td></tr>
<tr><td rowspan="2">Normative ethics (C3-2)</td><td rowspan="2">Informative (G) (C2)</td><td rowspan="2">Self-consciousness (C2)</td><td rowspan="2">Transformation and interpretation (epistemic)</td><td>Intrapersonal</td></tr>
<tr><td>Interpersonal</td></tr>
<tr><td>Reflexive ethics (C3-3)</td><td>Ethics (Ψ) (C3)</td><td>Moral consciousness (C3)</td><td>Synthesis and application (reflexive)</td><td>Existential/ethical</td></tr>
</table>

Table 4.2. *Interactions between levels of ethics, dimensions of systems modeling (Ψ, G, Φ) and forms of intelligence*

In short, it is the alliance of these three centers (of acquisition and organization, of transformation and interpretation and of synthesis and application) that makes possible the faculty of reasoning and moral deliberation at an articulated consciousness. Consequently, each entity or living organism is permanently in a situation of "acting and thinking" in complexity: thinking about its actions in organizational situations that it perceives in its constraints and opportunities in an evolving context. It is then essential to deploy the function "perceive, model" (informational rationality), and to launch the function "reason, deliberate" (cognitive rationality) without separating them, by launching these functions continuously in an organizational, interactive, collective and auto-eco-transformation of AI. The machine perceives "data", interprets "information", makes connections with other memorized "knowledge", and is then able to act with its "knowledge". It thus acquires skills that it is then able to implement repeatedly, translating "practical wisdom" or info-ethics (Ricœur 1990). This knowledge communication system (Todorova and Durisin 2007) is illustrated in Box 4.3.

Big Data issued → Simplified and relevant information received.

Recognize value → Acquire → Assimilate → Transform.

Information received → Knowledge.

Exploit (fueled by theory and experience) → Knowledge.

Knowledge → Practical wisdom (info-ethics).

Adaptations/flexibilities/innovations/performance through experience.

Box 4.3. *Knowledge communication processes*

By representing our complex knowledge through models, devices of symbols that we elaborate, through which we analyze and exchange, we manage to master this knowledge by making it simpler and, therefore, more intelligible. Symbolization is part of the purely formalistic tradition of the Pythagoreans. According to Simon and Newell (1971), the symbol device represents "the conjunction of form and relations, answering the question, 'What makes a symbol symbolize?' This search for practical wisdom contributes to a very low entropy (S^9), i.e. a degree of disorder that is almost nil".

In this way, we can understand the opening of knowledge in which humanity has been engaged since its origin. This process of ethical legitimization of the knowledge of the infosphere associated with the pragmatism of the environment of reality allows us to simplify our knowledge by approaching it both in a static way through its environment and in a dynamic way through its interactions. Under these conditions, ethics constitutes, on the one hand, the value characterized by absolute transcendence, and, on the other hand, the validity of pragmatic "immanence[10]". According to Morin (2004), ethics is inseparable from complex knowledge[11]: "knowledge that connects, that is expressed and built, as soon as one wants to be a responsible or supportive citizen". It is, therefore, essential to refocus our intelligence in order to face the complexity of the world around us and of ethics itself.

Finally, in order for representations to be unified, this unificatory power must be admitted as what allows knowledge, and therefore to think of it as originating. Consciousness is therefore an activity, a power of synthesis. The subject can only become aware of himself or herself through this activity. As the consciousness of the self appears only when it is realized, it cannot be a knowledge of oneself, because it is what allows knowledge. The subject can only be aware of something because he knows he is there – I am aware of the world only because I am aware of being there; world consciousness (C1) presupposes self-consciousness (C2). They are thus inscribed in an activity, in a movement, and therefore in a temporality that intertwines the primary consciousness (C1) and the reflexive one (C2) without making them coincide. Finally, the moral consciousness (C3) cannot reveal itself and function fully if it is not nourished by the two preceding levels of consciousness.

---

9 The word "entropy" was introduced by Clausius and is derived from a Greek word meaning "transformation". It always represents the complexity and, therefore, the possible disorder of a system, structure or organization.

10 "Immanence" is a philosophical term that designates the character of that which has its principle in itself.

11 The term "self-ethics" is used to refer to the need to "complicate judgment" in action. The word "self" means "fundamental autonomy" for the author.

4.4. Process of creating practical wisdom from artificial consciousness

Complex thinking loops together epistemology, anthropology and ethics. Epistemology allows us to conceive an anthropology, which is a primary condition for philosophical thinking, which is integrated into a loop where each step is necessary for the others to lead to an ethic. Finally, cybernetics will rely on info-ethics to develop AI. From these reflections, we can integrate our process of creating practical wisdom via the concept of organizational intelligence.

This ethical space of analysis takes up well the three levels of modeling of classical antiquity within the framework of a study of events, namely (see Figure 4.2):

– the "being given" represented by the epistemological aspect;

– the "sensitive world" illustrated by the anthropological field centered on human relations;

– the "objective reality" characterized by ethical and philosophical thinking. It consists first of all in respecting a deontology and human values so that the virtual integrates harmoniously with reality.

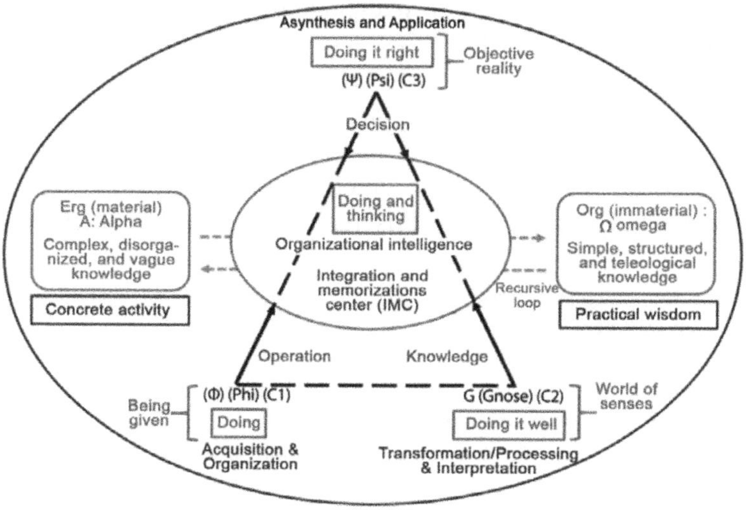

Figure 4.2. *Process of creating practical wisdom from an artificial consciousness. For a color version of this figure, see www.iste.co.uk/beranger/responsibility.zip*

The passage from concrete activity (alpha) to practical wisdom (omega) can also be characterized by the dialogical nature of the *erg (*material)[12] and the *org* (immaterial)[13] involved in the interaction that generates disorder and order. The organizational intelligence that enables this transformation integrates operation, knowledge and decision. The coherence of an organization constitutes the central element of the complexity of an event or a situation.

It seems necessary to move towards a "poetic[14]" knowledge that is meant to be heuristic, functional, endogenous – *sfumato* that connects and opens. This active reliance constitutes relearning to observe, elaborate, think and act. To constitute the recursive loop, always recharged with knowledge and reflections, of knowledge from knowledge. Thus, knowledge of knowledge requires complex thinking[15], which necessarily requires the intervention and reliance of the infosphere (informative dimension (G): knowledge), the environment of reality (environmental dimension (φ): operation) and info-ethics (ethical dimension (ψ): decision) constantly passing through each other.

This transition from alpha to omega necessarily involves the combination of pragmatics (acquisition and organization of data and information), epistemic knowledge (transformation and interpretation of information) and ethics (synthesis and application of knowledge) surrounding a decision. If the pragmatic aspect calls for ethical analysis, ethics refers to epistemological study, which calls for and activates reflection on the experience expressed by pragmatics[16]. Organizational intelligence of complexity links separate data, information and knowledge in order to move toward practical wisdom. This intelligence of reality represents a reconstitution and translation of this reality from a human mind. This is

12 The *erg* is the energetic activity of the system, permanently degeneration-regeneration involved in empowerment and greening.

13 The *org* represents the negentropic or informational activity of the system, generating in the organization by the energetic activity of the system, developing three functions: computation, information, memorization and communication.

14 The purpose of poetics is the study of the potentialities inscribed in a given situation that leads to a new creation.

15 Morin, E. (2006). L'esprit de reliance active l'organisation de la connaissance. Editorial Inter Lettre Chemin Faisant. *MCX-APC*, 35.

16 This can be illustrated by the viaticum of Leonardo da Vinci: "Sapience (science with knowledge) is the daughter of experience".

made possible by the Integration and Memorization Center (IMC): a meta principle representing an irradiating nucleus that centralizes, integrates, memorizes and connects triangulation (Ψ, G, Φ). The "perceptual tying" step is decisive in order to avoid the blinding problem. Indeed, the experiments go through separate sensory pathways. For example, we can describe a flower by its shape, its colors, its texture and/or its smell) and yet its perception is unique; it is a flower! The action of linking elements together is, therefore, decisive when we associate a perception and a concept. We link the vision of a flower to the semantic knowledge that we associate with a flower (its shape, its colors, its texture and/or its smell); we will also reconstruct memories (i.e. link pre-codified information to form structured and unified representations).

Thus, this IMC translating "doing" and "thinking" articulates and assimilates the "doing[17]" of the real environment, with the "doing it well" of the world of senses, with the "doing it right[18,19]" of objective reality.

The intelligibility of an organization's action has the capacity to maintain and sustain itself (self-regulation) and to link and relate (self-referencing) and to procreate and occur (autopoiesis). Thus, the formation of practical wisdom through organizational intelligence tends toward a decision-making system stemming from an artificial consciousness.

4.5. Morality of a "strong" AI

It is undeniable that new technologies, and in particular AI, must reflect moral and human values. Human beings will have to fight to preserve their creativity, their culture, their art and their social relationships that they have maintained and structured for millennia. The challenge of the 21st century is, therefore, to reinvent the place of human beings in the world, to rebuild in an intelligent way their way of living with artificial intelligent agents and their relationship with the planet. Thus, in the image of the seven deadly sins of the Christian tradition, we have listed seven pitfalls that future generations of "strong" self-learning, evolutionary and artificially conscious AI will have to avoid at all costs (Box 4.4).

---

17 To understand. In what context?

18 To do "why". For which projects?

19 To mobilize intelligence. For which decision?

Egocentrism

AI must not be programmed to survive in all circumstances, and especially not at the expense of its own operators and the human species.

Imposture by concealment

The machine must not behave in an opaque manner and/or pretend to be human, but must have an explainable and transparent reasoning and behavior toward its users in order to maintain trust and avoid betrayal to the users.

Naivety

The smart application must not be candid so that it is not a target for online scams or malicious acts. It must, therefore, have a high level of discernment in its human–machine and/or machine–machine interactions.

Cold perfectionism

AI should not announce results without qualification. It must show empathy by considering the situation.

Domination

The intelligent entity must not reject data or dismiss better options and decisions for humanity and the planet. It must not create dependencies through attention-grabbing techniques.

Narrow-mindedness

AI must not lead to a standardization of society through the standardization of behaviors and opinions. It must not limit and reduce the freedom to express ideas and share opinions, whose pluralism and diversity are one of the conditions of a democratic society.

Indiscretion

AI shall not disclose confidential and privacy information without the consent and authorization of the person and/or entity concerned.

Box 4.4. *Pitfalls to avoid for "strong" AI*

Is this future digital humanity preparing to live with other intelligences? Given the many debates and exchanges on this subject, we are convinced of this. In our opinion, it is precisely the anxieties and uncertainties about a cohabitation between the different organic and inorganic intelligences that arise in part from a current lack of knowledge about the evolution of intelligences (human, animal and artificial), their nature and their history, more precisely the co-evolution between the human species and its cultural and technical environments; hence the need to better understand these intelligences in all their diversity, similarity and difference. This is, in our opinion, an obligatory passage that we have set ourselves. It is then necessary for humans to free themselves from their superiority complex toward animals and their inferiority complex toward AI, to move away from a despairing vision of evolution, between scorned animals (out of contempt) and glorified robots (out of admiration or fear). Sterile arrogance and unproductive submission have never been great marks of intelligence. An evolutionary vision associating human, animal and artificial intelligences in sociocultural and technical environments is, therefore, obvious to us! Will such an approach stimulate human intelligence to better understand animal and artificial intelligence, or will it participate in a global dumbing down?

Our various research have revealed that human intelligence is partly structured in its relations to space and the environment, to body movements and to manipulations. We are convinced that AI as an intelligent machine building its own interactions forces us to rethink the anthropology of intelligences. This is why understanding animal intelligences and their interrelations with the environment is a *sine qua non* condition to understand future AI and their social roles in human civilizations. Consequently, animal intelligences with their modalities of appearance, evolution and diversification must be major sources of analogical bio-inspiration for AI. This necessarily requires the development of cobotics, i.e. the science that studies direct or remotely operated interactions and collaborations between humans – possibly including animals – and machines in non-artificial environments.

Conclusion

Technological developments have driven many changes in the way we interact, inform ourselves, work and think about the world. Therefore, the digital revolution that society is going through requires the acquisition of new skills. Technical advances and digital data processing represent a real asset, but also represent ethical risks. The growing awareness of the power of automatic decision-making systems, which are now omnipresent in all spheres of activity (aerospace, commercial, administrative, insurance, economic, industrial, medical, etc.), raises as many hopes as they do legitimate fears. One of the challenges of improving an AI algorithm is based on machine learning technologies, which allows it to learn from its mistakes. It is essential that algorithm designers are aware of the value judgments that impact their development. The way the algorithm is built is supported by values and imagined by people. For this reason, it is essential to list the different families of algorithms according to their purpose in order to better understand them, challenge them and adjust them to improve them. Indeed, Big Data may accentuate discrimination, bias, categorization, the risk of re-identification (hence the loss of confidentiality and privacy) and the individualization of society.

Data will increasingly associate with us, like a second skin, in order to better classify us and discriminate us with the drifts associated with it. This is why we should not underestimate the risk of a "digital divide" illustrated by several dividing lines (social, industrial, economic, even spatiotemporal) and covering multiple realities (see Appendix 6).

In this light, we need to rethink our approach to digital life, the use of AI and its impact on the environment. Protecting privacy requires users of the tools of this digital revolution to become more responsible for their actions. Individuals must be able to understand their digital environment in order to be in control of their applications. If a detailed ethical framework is not put in place, then the ethical glass ceiling will have exploded and we will not be able to change anything. The survival of the human species – as we now know it – will then be guided solely by the technical rationalization of the intelligent autonomous systems that are part of our ecosystem and for which we do not know the relevance.

In these conditions, it is necessary to provide a framework and ethical support that articulates the human and technical processes related to AI. Our digital ethics thus constitutes an adequate mode of regulation of human–machine–environment behaviors based on the adherence to moral values that we consider indispensable. It should not be considered as a constraint for the digital economy, but rather as a source of opportunities and fulfillment. Algorithms must be ethical and moral from their development to their use, since the responsibility belongs to both designers and owners. Thus, it is essential to bring a moral personality to these examples of AI through an ethical algorithmic consciousness that evolves over time. In our opinion, this notion of responsibility corresponds to the cornerstone of the framework around algorithmic systems; it is our approach of Ethics by Evolution!

On the other hand, the governance of digital data is a governance for the near future that calls for redefining the dimensions and aspects of today's ethics. Thus, it is the whole of our society that needs to define a digital ethic, namely scientific and expert institutions to define and apply common moral values, citizens who must become aware of the new digital challenges, digital professionals who must be oriented in their work as close as possible to action, and researchers in logic science (fundamental mathematics and theoretical computer science) to code the hierarchical complexity of ethical principles and human values from a dynamic logic system of neural network type; in other words, AI with an ethical vocation. As the French humanist writer of the Renaissance, François Rabelais, would have said so well, "Algorithms without consciousness are nothing but the announced ruin of our society". This is why ethics applied to AI fully concerns the future of the human species and of our freedom of determination, judgment and free will, both individually and collectively.

Moreover, it is crucial to have a holistic approach to digital. It is not just about technology and tools, but also about leadership, standards and rules.

Ethical AI also means how human beings interact with AI. We need empathy, education, creativity, judgment and responsibility.

It is from this context that this book is composed of, among others, 50 ethical actions associated with the 36 ethical criteria. Its major challenge is to provide an ethically and socially acceptable framework resulting from the development, design and launch of AI technologies in a digital project that can be illustrated by the drafting of an ethical charter. In addition, based on the identification of the main opportunities and risks of AI for society, as well as the set of four ethical principles that we have summarized to guide its development, we were able, first, to identify the challenges around the regulation of AI, and then, second, to establish rules of governance to create concrete and constructive responses to the most pressing social challenges posed by AI.

Therefore, we cannot rely exclusively on the responsibility of the protagonists of these changes nor on the dynamics of AI research to avoid slippage, drifts, or even the trivialization of abusive uses of these techniques. It is at the level where the majority of digital choices in everyday life are generated, at the individual level, that digital ethics are inscribed. This is why education about this new digital responsibility is essential. To do this, we need to socially impose the requirement of responsibility on citizens by abandoning an overly normative approach for a more ethical approach. Consequently, awareness and education campaigns will have to evolve in order to ensure that new generations are informed and trained to understand these technological innovations and the issues involved in the types of decisions associated with them as well as to use and secure their personal data. Therefore, our education system must make it a priority to transmit the knowledge and skills that will enable each individual to find his or her place in society, both as a future employee and as an active citizen.

AI used in an ethical way could mark an important, even primordial progress for humanity and the planet. Current investigations consist of seeking better compromises between different constraints: interpretability, explainability, integrity and quality of prediction, reduction of bias and drifts, and confidentiality of numerical data. To achieve this, auditing algorithmic processing via labeling, certification or even regulatory bodies seems to be essential. Currently, no single actor can claim to be able to control and guarantee the algorithmic loyalty of an eco-responsible and ethical AI. A plurality of counter-powers is, therefore, essential. Who are the protagonists likely to take charge of these regulations? Some are public regulators: CNIL, DGCCRF (fraud repression), competition authority, state jurisdictions, etc.; but do they have the capacity? Others are private – collaborative platforms (*Conseil national du numérique*, Data Transparency Lab, INRIA's Trans-Algo), media (ProPublica in the United States), data NGO (Bayes Impact), the company ORCAA[1] created by Cathy O'Neil, the start-up MAATHICS[2] and ADEL[3] – but are only at the beginning of their structural and economic development. To this, we can add initiatives to provide a deontological framework around developers, like the Holberton-Turing oath, which was generated with the aim of federating and raising the awareness of all professionals in the field of AI at the global level, around common moral and ethical values in order to invite them to use their skills with respect for humans while avoiding any threat to life (see Appendix 7), and the report of the French Parliamentary Office for the Evaluation of Scientific and Technological Choices, of March 15, 2017, entitled "*Pour une Intelligence Artificielle maitrisée, utile et démystifiée*" (see Appendix 8).

Finally, humanity is constantly being targeted by industrial revolutions that have sociocultural and economic repercussions and consequences on the world's population. These cyclical disruptions are part of an ineducable planetary Darwinian evolution. The next concerns are likely to focus on the integration of artificial consciousness ("strong" AI) into machines and the discoveries of quantum physics. In order to understand, prepare and support it as well as possible, our society will have to equip itself with an adapted moral and human framework by creating a new way of thinking, reflection, empowerment and a new way of conceiving value and work. From then on, we need to make sense of data, algorithms and their uses in order to bring real added value to the multifaceted AI that is invading our society and our environment. The difficulty will be to project the algorithmic system in its operation in relation to other autonomous systems; in other words, to control the consequences of its future performances that will be made possible by future quantum computers. Under these conditions, it will be necessary to elaborate, construct and develop a true quantum ethic. Because, in our opinion, writing and thinking about the future of an algorithmic and then quantitative ethic is simply making it possible! In our opinion, the future of humanity lies in its capacity to preserve, on the one hand, human thought

1 http://www.oneilrisk.com/.

2 http://www.adel-label.com/.

3 http://fdu-label.com/fr/.

and consciousness, and, on the other hand, inter-human communication. This will probably pass by a progressive digitalization of all information, reflections, even in the medium term of human consciousness integrating the notion of Ethics by Evolution.

Appendices

Appendix 1

Ethical Charter of Using AI in Judicial Systems and Their Environment

1) Principle of respect for fundamental rights: It ensures the design and implementation of AI tools and services that are compatible with fundamental rights.

2) Principle of non-discrimination: It specifically prevents the creation or reinforcement of discrimination between individuals or groups of individuals.

3) Principle of quality and security: With regard to the processing of jurisdictional decisions and forensic data, we use certified sources and intangible data with models designed in a multidisciplinary manner, in a secure technological environment.

4) Principle of transparency, neutrality and intellectual integrity: It makes data processing methodologies accessible and understandable, and authorizes external audits.

5) Principle of control by the user: It banishes a prescriptive approach and allows the user to be an enlightened actor and master of their choices.

Appendix 1

Ethical Charter of Using AI in Judicial Systems and Their Environment

1) Principle of respect for fundamental rights: It ensures the design and implementation of AI tools and services that are compatible with fundamental rights.

2) Principle of non-discrimination: It specifically prevents the creation or reinforcement of discrimination between individuals or groups of individuals.

3) Principle of quality and security: With regard to the processing of jurisdictional decisions and forensic data, we use certified sources and intangible data with models designed in a multidisciplinary manner, in a secure technological environment.

4) Principle of transparency, neutrality and intellectual integrity: It makes data processing methodologies accessible and understandable, and authorizes external audits.

5) Principle of control by the user: It banishes a prescriptive approach and allows the user to be an enlightened actor and master of their choices.

Appendix 2

Practical Recommendations of the CNIL Regarding the Ethics of Algorithms

1) *Train all the actors involved in the "algorithmic chain" in ethics*: designers, professionals, citizens. Digital literacy must enable every human being to understand the workings of the machine.

2) *Make algorithmic systems understandable* by strengthening existing rights and organizing mediation with users.

3) *Work on the design of algorithmic systems in the* service of human freedom to counter in particular the "black box" effect.

4) *Constitute a national platform for auditing algorithms.* This is essential to reinforce the confidence of all actors in the health system.

5) *Encourage the search for technical solutions to make France the leader in ethical AI.* Clearly, initiatives such as the constitution of a future national health data "hub" will need to put ethical considerations at the center of the project.

6) *Strengthen the ethics-related roles within companies.* Ethics is also a competitive argument that must be based on the development of ethics committees, the dissemination of good industry practices and the development or revision of charters of deontology.

Appendix 3

OECD Recommendation on AI

The recommendation is divided into two main sections:

– principles of a responsible approach to support trusted AI: this first section sets out five complementary principles of interest to all stakeholders: (1) inclusive growth, sustainable development and well-being; (2) people-centered values and equity; (3) transparency and accountability; (4) robustness, safety and security; and (5) accountability. It further calls on AI stakeholders to promote and implement these five principles, according to their respective roles;

– national policies and international cooperation in support of trusted AI in line with the five principles mentioned above: this second section sets out five recommendations for members and non-members who have adhered to the draft recommendation (hereinafter referred to as "adherents"), which they are invited to implement in the framework of their national policies and international cooperation: (1) invest in AI research and development; (2) foster the development of a digital ecosystem for AI; (3) shape a supportive policy framework for AI; (4) build human capacity and prepare for the transformation of the labor market; and (5) foster international cooperation for trusted AI.

Appendix 4

Questions Concerning the Application of Ethical Standards

A4.1. Question 1: universality of standards

– Concerns: The standards to be integrated in AI are not universal, but specific to the actions and nature of the latter and the company that will use it.

– Recommendations are as follows:

- identify the ethical standards that AI must follow according to its purposes and its actions with the society;
- set up empirical research that involves several disciplines and methods to study and document these ethical standards;
- generalize, at the national and then international level, multi-disciplinary ethics committees in engineering sciences with common ethical standards.

A4.2. Question 2: moral saturation

– Concerns: AI is subject to a moral overload due to a multiplicity of norms that may conflict with each other.

– Recommendations are as follows:

- prioritize ethical standards that best reflect the set of values shared by stakeholders;
- prioritize ethical standards from the design of AI;

- establish an explicit justification system for the decision and choice of a standard associated with one value over another[1];

- integrate an interactive machine learning device, or questioning and modeling of user responses to capture the changing ethical rules of users over time and in different contexts.

A4.3. Question 3: bias

– Concerns: IAs can be the target of construction and/or operation bias.

– Recommendations are as follows:

- identify and understand all potential and conceivable biases associated with AI;

- carry out multidisciplinary research to better understand the nature of these biases and their alignment with human and moral values;

- study and become familiar with the specific resources and characteristics (expectations, concerns, vulnerabilities, human values, etc.) of the populations[2] that will be the target of AI;

- include regulators and policy makers in the AI development process in order to shape regulations adapted to the target populations associated with AI.

A4.4. Question 4: integration and compatibility of standards

– Concerns: How can ethical standards be properly integrated into a computing architecture? For example, which algorithm should I program if I need to avoid confusion between an elderly person or a child?

– Recommendations are as follows:

- develop research that addresses how to incorporate ethical standards into AI (data collection, sensor technology, pattern recognition, machine learning and integration of different types of data, etc.);

---

1 This can have the effect of encouraging designers to think more about the values and standards to be incorporated into AI, but it can also provide a basis and reference point for a third party to better understand the thinking process of the designer.

2 We note that vulnerable populations are often one of the first targets of AI users.

Appendix 6

Reasons for a "Digital Divide"

The reasons for a digital divide are as follows:

– The complexity of understanding namely when, why and how processed digital data were prioritized and placed in a specific category. This is essential in order to reinforce their self-control.

– The inability to access the methodology and individual know-how, the often opaque and hidden operation of the algorithm, and the decision criteria that were applied.

– The difficulty for individuals to be able to process and adjust digital data.

– The ability or not for the persons concerned to be informed about the traceability of the data throughout its lifecycle.

– The complexity of access to NICTs (access to a suitable machine, connection to the network, bearable costs, etc.), as well as to content (i.e. the ability to truly process the technical, informational and communicational resources that these technologies contain).

– The difficulty for individuals and structures to access or purchase algorithmic applications.

- know what you have done and what you want to do or not do when designing AI. It is not up to AI to make choices for the company, but to the human being (via the coder and/or the developer) to decide what he or she will bring to it;

- make standards (machine; human) compatible according to context and target user populations;

- evaluate by empirical means[3] the success of the implementation of ethical norms in AI throughout the development cycle of the autonomous system.

A4.5. Question 5: trust

– Concerns: People's anxiety about the impact of IAs is growing, which contributes to a decrease in trust in IAs.

– Recommendations are as follows:

- implement a module within AI that guarantees a certain degree of transparency[4] and reliability of AI;

- communicate the limitations and capabilities of AI to users;

- install in AI a module of functionalities that prevents users from operating it according to the conditions of use, some of which depend on the behaviors;

- evaluate the purpose, behavioral characteristics of AI and its interaction with the user;

- develop a monitoring, traceability and recording system (of the "black box" type) associated with changes, or even malfunctions (decisions, behaviors, etc.) of the AI;

---

3 These means may correspond to in-depth testing (including contradictory tests), capacities to explain the functioning of autonomous systems, natural language dialogue between AI and humans (including response) and consciousness of the context and memory (to manage repeated evaluations).

4 This transparency, synonymous with confidence in the algorithm, is a major strategic issue for the company. It will highlight the value judgment of its designer, and, therefore, of the company that operates it, giving it a major competitive advantage in a market where customer loyalty has become a must.

- assess by a trusted third party the alignment of human values, conformity and appropriate capabilities with the actions and decisions of AI;

- define evaluation criteria (reference scale[5]) to measure the level of ethics of AI;

- introduce new forms of codes of conduct for coders, developers and even users;

- mechanisms for certification and auditing of automated data processing techniques like algorithms should be developed to ensure that these techniques respect human rights. Public entities and non-state actors should encourage and promote the strengthening of human rights through ethical design strategies and stronger risk assessment procedures in software development.

A4.6. Question 6: environment

– Concerns: The development and operation of AI requires a very large amount of energy. Therefore, these NICTs consume a lot of energy.

– Recommendations are as follows:

- think about the human–machine–environment triptych rather than only the human–machine relationship;

- develop AI for the benefit of humans as well as for the benefit of the planet;

- install a system within NICTs that emits a large amount of heat, a system that allows this surplus heat to be reused for ecological purposes to avoid wasting energy (digital ecology);

- set up research that considers the stakes and issues surrounding the digital environment;

- create a scientific collaborative platform, intended to promote the development of ecological AI.

5 These criteria will depend in part on the expected tasks of the machine and the context of use, as well as on user vulnerabilities. It is desirable that the more vulnerable categories of users require stricter criteria.

Appendix 5

CERNA Recommendations on Machine Learning

CERNA recommendations on machine learning are as follows:

1) Designers and coaches of learning systems ensure the quality of the learning data and the conditions of their capture.

2) Trainers must ensure that the data reflects cultural diversity.

3) Variables for which data are regulated, and coaches must ensure that they are non-discriminatory (age, gender, race, etc.) while respecting the principle of data confidentiality.

4) The designer of a machine learning system shall provide system traceability features.

5) The machine must not introduce any characterization bias and mislead the user on the state of the system.

6) The designer must maintain a certain level of vigilance in communicating the capabilities of a learning system in order to leave no room for misinterpretation or irrational fantasies or fears.

7) The designer must ensure the explainability and the transparency of the actions of the learning system, while maintaining a sufficient level of performance.

8) While ensuring a better explainability of the system, the designer must describe the limitations of the system's explanatory heuristics, in particular avoiding the creation of biases.

9) The developer of a learning system contributes to the development of standards and protocols for evaluating machine learning.

10) The designer must guarantee the place of the human in the decisions assisted by learning machines in order to avoid, in particular, the creation of bias or the installation of dependence of the human in relation to the decisions of the machines.

11) The storage of traces of personal data used in the learning process shall require the consent of the user in accordance with the applicable personal data protection legislation.

12) The system designer shall include automated or supervised control mechanisms.

13) The designer must provide a statement of the intentions to use the computer system "sincerely, faithfully and completely" during the learning process.

14) The creation of a national research network called the "Federative Initiative for Digital, Ethical, and Social Research" would make it possible to develop a national position on the questions of the societal and ethical impact of digital sciences and technologies.

15) The creation of operational ethics committees of digital science and technology institutions is advised.

16) Institutions are also encouraged to launch initiatives on the legal aspects of the uses of digital innovations, through working groups and research projects with other stakeholders.

17) Actions to raise awareness and support the researcher by the institutions must be put in place.

Appendix 7

Holberton–Turing Oath

The Holberton–Turing oath is based on three main principles: respect for life and science, inclusion and the transmission of knowledge. The principles of the Holberton–Turing oath will be able to fuel discussions on subjects such as respect for personal data, the development of autonomous weapons, the loss of jobs by tens of millions and the creation of technological discrimination. This oath has been built in the image of an open-source code, intended to evolve over time according to the contributions of each player on the technological and social chessboard. Finally, it has been designed to be understood by all peers in the discipline, as well as by citizens, who will see, in the existence of this text, a basis for reflection to be extended and deepened in public debates.

A7.1. Holberton–Turing oath

At the moment of being admitted to exercise artificial intelligence in all its present and future forms, I promise and swear to be faithful to the laws of honor and probity.

A7.1.1. *Humanity and ethics*

– I will preserve the human being and work to restore, preserve or promote equity and ethics in all its elements, physical and mental, individual and social.

– I will respect all persons, their autonomy and will, without any discrimination on the basis of age, physical condition, political affiliations, religious beliefs, social origins, ethnic origins or sexual orientation.

– I will never mislead their trust and will not exploit the power inherited from circumstances to force consciences.

A7.1.2. *Data science, the art of artificial intelligence, privacy and personal data*

– I will respect the hard-won scientific advances and progress made by the scientists and engineers who preceded me, and will share the knowledge I possess with those who follow me.

– I will remember that there is an art to artificial intelligence, as well as to science, and that human concerns outweigh technological concerns.

– I will respect the privacy of users and ensure that their personal data are not disclosed.

– I will remember that I do not just manipulate data – zeros and ones – but human beings whose interactions with my artificial intelligence software can affect freedom, family and economic stability.

– I will respect the secrets entrusted to me.

A7.1.3. *Daily work and etiquette*

– I will practice my profession with conscience and dignity.

– I will respect and pass on the honor and noble traditions of the profession of data science and artificial intelligence.

– I will give my professors, colleagues and students the respect and gratitude they deserve.

– I will share my knowledge for the benefit of the greatest number and the advancement of the science of data and artificial intelligence.

– I will examine the impact of my work on equity both in the perpetuation of historical biases, which can be caused by the blind use of past data to future predictions, and in the creation of new conditions that increase economic or other forms of inequality.

– I promise to create software solutions that incorporate artificial intelligence, focusing on working with humans for the greater good, rather than usurping and supplanting the role of humans.

I make these promises solemnly, freely and on my honor.

Appendix 8

Report Proposals: "For a Controlled, Useful and Demystified Artificial Intelligence"

A8.1. For a controlled artificial intelligence

Proposal no. 1: Avoid too strong a legal constraint on research in artificial intelligence (AI), which, in any case, would benefit from being, as much as possible, European, or even international, rather than national.

Proposal no. 2: Promote safe, transparent and fair algorithms and robots and provide a charter for AI and robotics.

Proposal no. 3: Training in AI and robotics ethics in some specialized higher education courses.

Proposal no. 4: To entrust a national institute of AI and robotics ethics with the role of leading the public debate on the main ethics that should govern these technologies.

Proposal no. 5: To support the transformations of the labor market under the effect of AI and robotics by carrying out an ambitious continuous training policy aimed at adapting to the requirements of requalification and improvement of skills.

A8.2. For useful AI, in the service of humans and humanistic values

Proposal no. 6: Restore an essential place to basic research and revalue the place of public research in relation to private research while encouraging their cooperation.

Proposal no. 7: Encourage the development of European champions in AI and robotics.

Proposal no. 8: Direct investment in AI research toward the social utility of discoveries.

Proposal no. 9: Broaden the offer of courses and training modules in AI technologies in higher education and create – in each nation – at least one international and interdisciplinary center of excellence in AI and robotics.

Proposal no. 10: Structure and mobilize the global AI research community by organizing more award-winning competitions on a national scale, aimed at boosting AI research, for example, around the processing of large national labeled databases.

Proposal no. 11: Ensure better consideration of the diversity and place of women in AI research.

A8.3. For a demystified AI

Proposal no. 12: Organize computer training in primary and secondary education with a focus on AI and robotics.

Proposal no. 13: Train and raise public awareness of AI through communication campaigns, the organization of an international exhibition of AI and robotics, and the broadcasting of educational television programs.

Proposal no. 14: Train and raise awareness of the general public on the practical consequences of AI and robotics.

Proposal no. 15: Be vigilant on the spectacular and alarmist uses of the AI concept and representations of robots.

List of Abbreviations

A2B	Administration to business
ACC	Accessibility
ACC	Accountability
ADA	Adaptability
ADEL	Algorithm Data Ethics Label
AI	Artificial intelligence
ANN	Announcement
APP	*Agence pour la protection* [French Agency for Program Protection]
APP	Applicability
AUT	Autonomization/autonomy/automation
B2B	Business to business
BI	Business intelligence
BIA	Bias

CCNE	*Conseil consultatif national d'éthique* [French National Consultative Ethics Council]
CDE	Chief Digital Ethics
CEPEJ	European Commission for the Efficiency of Justice
CERNA	*Commission de réflexion sur l'éthique de la recherche en sciences et technologies du numérique d'Allistene* [Reflective commission on the ethics of research in science and Allistene's digital technologies]
CFO	Chief Financial Officer
CNIL	*Commission nationale de l'informatique et des libertés* [National Commission on Informatics and Liberty]
COL	Non-selective collection
CON	Confidentiality
CON	Consistency
COP	Conference of the Parties
CSA	*Conseil supérieur de l'audiovisuel* [Superior Audiovisual Council]
CUL	Culture
DE	Data ethics
DEO	Deontology
DEH	Dehumanization
DGCCRF	*Direction générale de la concurrence, de la consommation et de la répression des fraudes* [Directorate-General for Competition, Consumer Affairs and Fraud Control]
EA	Ethics of algorithms

ED	Ethics of decisions
ENV	Environment
EP	Ethics of practices
ERG	Ergonomics
ES	Ethics of systems
EU	European Union
EXP	Explainability
FWI	Free will
GDPR	General Data Protection Regulation
GOV	Governance
HPC	High-performance computer
HR	Human resources
HSS	Human and social sciences
IEEE	Institute of Electrical and Electronics Engineers
IETF	Internet Engineering Task Force
INC	Inclusion
INRIA	*Institut national de recherche en informatique et en automatique* [French Institute for Research in Computer Science and Automation]
INT	Integrity
ISO	International Organization for Standardization
IT	Information technology

JORF	*Journal officiel de la République française* [Official Journal of the French Republic]
LGBTQIA+	Lesbian, gay, bisexual, transgender, queer, intersex and asexual people
MAN	Management
NBIC	Nanotechnologies, biotechnologies, computer and cognitive sciences
NICT	New information and communication technologies
NGO	Non-governmental organization
ORCAA	O'Neil Risk Consulting and Algorithmic Auditing, Inc.
ORG	Organization
PER	Performance
PLC	Programmable logic controllers
PRI	Privacy
PRO	Protection
PUR	Purpose
QUA	Quality
R&D	Research and development
REG	Regulation
REL	Reliability
RES	Responsibility
ROI	Return on investment
SEC	Security

SQL	Structured Query Language
SWOT	Strengths, weaknesses, opportunities and threats
TRA	Traceability
TRAN	Transparency
TRU	Trustworthiness
UEM	Urban environmental monitoring
Wi-Fi	Wireless Fidelity
WWF	World Wide Fund for Nature

References

Alexandre, L. (2017). *La guerre des intelligences*. JC Lattès, Paris.

Amann, J.P. and Gaille, M. (2007). Approche par les principes, approche par les cas : les limites philosophiques d'une opposition. *Éthique et santé*, 4, 195–199.

Balagué, C. (2017). De l'éthique des algorithmes et de l'intelligence artificielle. In *Vade-mecum sur la gouvernance des traitements de données numériques*. ADEL, Paris.

Barocas, S., Andrew, D., Selbst, A.D. (2016). Big Data's disparate impact. *California Law Review*, 104, 671.

Beauchamp, T.L. and Childress, J. (2001). *Principles of Biomedical Ethics*, 5th edition. Oxford University Press, Oxford.

Bensamoun, A. and Loiseau, G. (2017). L'intelligence artificielle à la mode éthique. *Recueil Dalloz*, 1371–1372.

Béranger, J. (2015). *Medical Information Systems Ethics*. ISTE Ltd, London, and John Wiley & Sons, New York.

Béranger, J. (2016). *Big Data and Ethics: The Medical Datasphere*. ISTE Press, London, and Elsevier, Oxford.

Boisvert, Y., Jutras, M., Legault, G.A., Marchildon, A. (2003). Raisonnement éthique dans un contexte de marge de manœuvre accrue : clarification conceptuelle et aide à la décision. Research report, Gouvernement du Québec/Centre d'expertise en gestion des ressources humaines, Quebec.

Bollier, D. (2010). *The Promise and Peril of Big Data*. The Aspen Institute, Washington.

Bonhomme, D. and Pinaudeau, D. (2008). Imagerie médicale : des innovations au service de l'optimisation et de l'efficience. État de l'art en imagerie. *IRBM News*, 20124.

Davis, K. and Patterson, D. (2012). *Ethics of Big Data: Balancing Risk and Innovation*. O'Reilly Media, Sebastopol.

De Ganay, C. and Gillot, D. (2017). Pour une intelligence artificielle maîtrisée, utile et démystifiée. Report, Office parlementaire d'évaluation des choix scientifiques et technologiques, Paris.

Dherse, J.-L. and Minguet, D.H. (2007). *L'éthique ou le chaos ?* Presses de la Renaissance, Paris.

Fessler, J.-M. and Grémy, F. (2001). Ethical problems in health information systems. *Methods Inf. Med.*, 40(4), 359–361.

Floridi, L., Cowls, J., Beltrametti, M., Chatila, R., Chazerand, P., Dignum, V., Luetge, C., Madelin, R., Pagallo, U., Rossi, F., Schafer, B., Valcke, P., Vayena, E. (2018). AI4People – An ethical framework for a good AI society: Opportunities, risks, principles, and recommendations. *Minds and Machines*, 28(4), December.

Fransman, M. (1994). Information, knowledge, vision and theories of the firm. *Industrial and Corporate Change*, 3, 1–45.

Gruson, D. (2018). *La Machine, le médecin et moi*. Éditions de l'Observatoire/ Humensis, Paris.

Gruson, D. (2019). Intelligence artificielle et santé : quoi de neuf docteur ? *Institut Montaigne* [Online]. Available at: https://www.institutmontaigne.org/publications/ ia-et-emploi-en-sante-quoi-de-neuf-docteur.

Gu, L., Kropotov, V., Fyodor Yarochkin, F. (2017). The fake news machine: How propagandists abuse the Internet and manipulate the public. Trendlabs research paper, Trend Micro, Tokyo.

Guttman, N. (1996). *Public Health Communication Interventions. Values and Ethical Dilemmas*. Sage Publications, Thousand Oaks.

Harari, Y.N. (2017). *Homo deus : une brève histoire de l'avenir*. Albin Michel, Paris.

Hervé, C. (1997). *Éthique médicale ou biomédicale ? L'éthique en mouvement*. L'Harmattan, Paris.

Höffe, O. (1991). *La justice politique : fondement d'une philosophie critique du droit et de l'État*. PUF, Paris.

Kepeklian, G. and Wibaux, G. (2012). Quels modèles économiques pour le Big Data ? *ATOS*, July, 1–54.

Khare, R., Good, B.M., Leaman, R., Khare, R., Good, B.M., Leaman, R., Su, A.I., Lu, Z. (2015). Crowdsourcing in biomedicine: Challenges and opportunities. *Briefings in Bioinformatics*, 17(1), 23–32.

Le Coz, P. (2007). *Petit traité de la décision médicale*. Le Seuil, Paris.

Le Coz, P. (2009). Place de la réflexion philosophique dans la décision médicale. *Bulletin de l'académie nationale de médecine*, 193(2), 499–510.

Le Moigne, J.L. (1999). *La modélisation des systèmes complexes*. Dunod, Paris.

Marcinkowski, A. and Wilgaux, J. (2004). Automates et créatures artificielles d'Héphaïstos : entre science et fiction. *Techniques & culture*, 43–44, 9.

Morin, E. (2004). *La Méthode 6 : éthique*. Le Seuil, Paris.

Morin, E. (2005). Complexité restreinte, complexité générale. In *Colloque Intelligence de la complexité : épistémologie et pragmatique*, Le Moigne, J.-L. and Morin, E. (eds). Cerisy-La-Salle.

Nye, D.E. (2008). *Technologie & civilisation : 10 questions fondamentales liées aux technologies*. FYP, Limoges.

Ogien, R. (2007). *L'éthique aujourd'hui : maximalistes et minimalistes*. Gallimard, Paris.

Ponçon, G. (2009). Système d'information : vers une éthique de l'usage ? *Revue hospitalière de France*, 531, 74–77.

Picard, F. (2017). Une éthique du numérique centrée sur les citoyens et orientée vers des solutions techniques. In *Vade-mecum sur la gouvernance des traitements de données numériques*. ADEL, Paris.

Rastier, F. (2001). *Sémantique et recherches cognitives*. PUF, Paris.

Roth, F. (2004). Des mécanismes de gouvernance informationnels aux mécanismes de gouvernance cognitifs : une grille de lecture et quelques conséquences pour la communication financière des entreprises. In *Congrès gouvernance et juriscomptabilité*. HEC Montréal, Montreal.

Riach, P.A. and Rich, J. (2002). Field experiments of discrimination in the market place. *The Economic Journal*, 112(483), 480–518.

Ricœur, P. (1990). *Soi-même comme un autre*. Le Seuil, Paris.

Ricœur, P. (1991). Éthique et morale. In *Lectures 1 : autour du politique*. Le Seuil, Paris.

Shen, Y., Colloc, J., Jacquet-Andrieu, A., Lei, K. (2015). Emerging medical informatics with case-based reasoning for aiding clinical decision in multi-agent system. *Journal of Biomedical Informatics*, 56, 307–317.

Shen, Y., Yuan, K., Chen, D., Colloc, J., Yang, M., Li, Y., Lei, K. (2018). An ontology-driven clinical decision support system (IDDAP) for infectious disease diagnosis and antibiotic prescription. *Artificial Intelligence in Medicine*, 86, 20–32.

Simon, H.A. and Newell, A. (1971). Human problem solving: The state of the theory in 1970. *American Psychologist*, 26(2), 145–159.

Terry, N. (2014). Health privacy is difficult but not impossible in a post-HIPAA data-driven world. *Chest*, 146(3), 835–840.

Todorova, G. and Durisin, B. (2007). Absorptive capacity: Valuing a reconceptualization. *Academy of Management Review*, 32(3), 776.

Valéry, P. (1941). *Les carnets de Léonard de Vinci*. Gallimard, Paris.

Valéry, P. (1948). *Vues*. La Table Ronde, Paris.

Vayre, J.-S. (2018). Élément de réflexion éthique sur la socialisation algorithmique. In *Vade-mecum sur la gouvernance des traitements de données numériques*. ADEL, Paris.

Villani, C. (2018). Donner un sens à l'intelligence artificielle : pour une stratégie nationale et européenne. Report, Rapport de la mission confiée par le Premier Ministre, 242.

Williatte, L. (2017). Le cadre légal et réglementaire : les exigences du règlement européen du 27 avril 2016 et la loi française pour une république numérique du 7 octobre 2016. In *Vade-mecum sur la gouvernance des traitements de données numériques*. ADEL, Paris.

Zolynski, C. (2015). Big Data : pour une éthique des données. *I2D*, 2, 25–26.

Zouarhi, S. (2017). Éthique et TIC, la Blockchain comme outil de gouvernance des données. In *Vade-mecum sur la gouvernance des traitements de données numériques*. ADEL, Paris.

Index

A, B

C

D, E

F, H, I, J

M, N

P

Q, R

S

T, U

Other titles from

iSTE

in

Health Engineering and Society

2020

BARBET Jacques, FOUCQUIER Adrien, THOMAS Yves
Therapeutic Progress in Oncology: Towards a Revolution in Cancer Therapy?
(Health and Patients Set – Volume 2)

MORILLON Laurent, CLAVIER Viviane
Health Research Practices in a Digital Context
(Health Information Set – Volume 4)

NADOT Michel
Discipline of Nursing: Three-time Knowledge

PAILLIART Isabelle
New Territories in Health
(Health Information Set – Volume 3)

2019

CLAVIER Viviane, DE OLIVEIRA Jean-Philippe
Food and Health: Actor Strategies in Information and Communication (Health Information Set – Volume 2)

PIZON Frank
Health Education and Prevention (Health and Patients Set – Volume 1)

2018

HUARD Pierre
The Management of Chronic Diseases: Organizational Innovation and Efficiency

PAGANELLI Céline
Confidence and Legitimacy in Health Information and Communication (Health Information Set – Volume 1)

2017

PICARD Robert
Co-design in Living Labs for Healthcare and Independent Living: Concepts, Methods and Tools

2015

BÉRANGER Jérôme
Medical Information Systems Ethics

Printed and bound by CPI Group (UK) Ltd, Croydon, CR0 4YY

15/09/2026

14962361-0001